Casa de Hacienda

Architecture in the Colombian countryside

Casa de Hacienda

Architecture in the Colombian countryside

Director, designer and editor
BENJAMIN VILLEGAS

Texts
GERMAN TELLEZ

Photography
ANTONIO CASTAÑEDA

Villegas editores

This book has been created, produced and published
in Colombia by
©VILLEGAS EDITORES 1997
Avenida 82 No. 11-50, Interior 3
e-mail villegas@colomsat. net.co
Telephone (57-1) 616 1788. Fax 616 0020, 616 0073
Bogotá D.C., Colombia

Editorial consultant
FRANCA PACINI

Layout
MERCEDES CEDEÑO

Graphic coordination
PILAR GOMEZ

English translation
JIMMY WEISKOPF

First Edition
November, 1997

ISBN
958-9393-34-9

The publisher wishes to express special thanks to
DAVIVIENDA
for sponsoring the first edition of this book.

Jacket, *Santa Bárbara,* Tibitó, Cundinamarca.
Back jacket, *El Paraíso,* Valle del Cauca.
Page 1, *El Salitre,* Sopó, Cundinamarca.
Page 2, *Fagua,* Cajicá, Cundinamarca.
Pages 4/5, *Fagua,* Cajicá, Cundinamarca.

Téllez, Germán, 1933-
Casa de hacienda : architecture in the colombian countryside / text Germán Téllez ; photography Antonio Castañeda. -- Director, designer and editor Benjamín Villegas. -- Bogotá : Villegas Editores, 1997.
304 p. : il. ; 28 cm.
Incluye bibliografía e índice.
ISBN 958-9393-34-9
1. Arquitectura colombiana 2. Arquitectura doméstica - Colombia 2. Construcciones rurales - Colombia I. Castañeda, Antonio, il. II. Villegas, Benjamín, 1948- , ed. III. Tít.
720.9861 cd 19 ed.
AFY9282

Foreword

If the history of the Spanish conquest and colonization of the New World is that of the seizure of a vast territory, the expansion of an empire, the subjugation of native peoples who fiercely resisted the invaders and the imposition of one culture upon another is also the history of the development, in that territory, of the heritage and civilization which were brought by the Spanish. The colonial epoch witnessed a prolonged process of settlement and social organization marked by establishment of towns, building of churches, promulgation of laws, settling of farms, the sharing out of properties worked by indentured native labour and the awarding of royal land grants.

It is in this context that the first hacienda houses, with their characteristic architecture, arose in Nueva Granada, the original Spanish colonial name for what is now the territory of Colombia.

During the sixteenth and seventeenth centuries, when the post-Conquest landowners took possession of their dominions in Nueva Granada, this architecture derived directly from Spanish models. The first hacienda houses appeared on properties which were centers of power and were intended to be self-sufficient. The remote origin of this architecture is found in ancient Greece and Rome, but it reached its peak in the Spanish provinces of Andalusia, Granada and Extremadura, the point of departure for the great expeditions that eventually led to the Spanish settlement of Nueva Granada. As a result the hacienda houses of the new colony were directly related to the Islamic architecture in Spain built during the long period of Arabic domination. Their general framework, maintained a basic architectural and spatial order that was common to all the regions and climates of Nueva Granada, as Germán Téllez explains in this book.

It is clear that the rural Neo-Granadian house included many of the basic attributes of Hispanic ones, especially the "sense of place", that is, a rational benefits which may be derived from the site for the comfort and well-being of its inhabitants. These practical considerations were complemented by a no less important dimension: the warmth and grace of a shelter that offers a sure expansiveness to the spirit. In his moments of rest and intimacy the inhabitant of a hacienda house finds refuge within an architecture that belongs to a noble tradition.

The hacienda house finds its identity in its relationship with two of its fundamental aspects: the interior and the exterior.

The first consists of comfortable austere but spaces that are spacious but never ostentatious. The second has to do with the integration with a surrounding landscape that acts as its "external form" and makes the experience of the building and its surroundings an inseparable one. Along with ideas derived from established technical, stylistic and formal patterns – and in the absence of architects – an air of creativity prevails that is intuitive and logical. With the help of artisans and masons, the landowners of that period became the builders of their own houses, moulding the spaces to their own needs, on the basis of a tacit sense of constructional design.

From the second half of the nineteenth century onwards the triumphant ideas of the republican period began to superimpose themselves on a good part of colonial architecture and the hacienda house was not an exception. While some traditional features were conserved, the imposition of new elements often hid or destroyed a valuable architectural heritage. Even so, it was only much later, in the nineteen-eighties, that the architecture of the hacienda house began to dramatically lose its vital signs of identity.

As a result, a great cultural continuity has been disrupted: architecture, whose traditional values give a sense of cultural heritage, has been losing its solidity and the immediacy of modernity erases profound traces of the past. The danger that this architecture may lose its integrity is a constant threat and a sad reflection upon our culture… or lack of one.

BENJAMIN VILLEGAS

Opposite page, *Hatoviejo,* Yotoco, Cauca Valley.

Introductory note

[A hacienda may be roughly defined as a country estate: for a more precise definition, see the following chapter. Translations of Spanish book and article titles are given in the bibliography. Nueva Granada is the name for the Spanish colonial territory of what is now Colombia, which also embraced parts of present-day Ecuador and Venezuela: as an adjective for Nueva Granada, the term Neo-Granadian is employed].

In 1975, in one of the chapters of la Historia del Arte en Colombia, the author of this essay noted that:

"The building of a hacienda house materializes and awards a symbolic quality to the relation of man with the countryside. The land requires the person who possesses it to provide a permanent testimony of his affection for it and so, he who builds a house on properties which he believes are his own is the final and definitive owner of them. It is in the country that the Spanish terms 'dueño' [landlord] and 'Señor' [lord, master] acquire their true meaning. It was literally through their rural architecture that the Spanish peoples who reached the lands which they called Nueva Granada put down deep and fertile roots."

The basic subject of this book is the relationship of house-place-landscape, treated from a point of view that is directly opposite to that of conventional writers and chroniclers, that is, *starting* with the presence of the forms constructed in the natural space. Even granting the obvious pre-existence of the countryside as a setting, rural architecture is a real, solid and conceptual basis for establishing a fixed perceptive approach to it. In accordance with this premise, landscape is that which is defined by the presence of architecture, *from the moment in which it makes its appearance.*

The uninhabited countryside, the rugged or desolate landscape, possesses a separate kind of natural order, or, if you like, a certain chaos prior to architecture, endowed with meanings and values for the human being which radically change when he decides to leave his mark on it. To support the above criterion we may refer to the work of the painters and draughtsmen of colonial Nueva Granada and republican Colombia from the nineteenth up to the beginnings of the twentieth centuries, in which, contrary to what happens in literature, the constructional forms of the countryside become an object of increasing interest, either as component elements of their treatment of the observed landscape or as the primary subject matter.

The sense of place, together with the human being's awareness of vital space, has led him, throughout history, to act as an animal ("The only animal which knows it has to die," according to André Malraux) that biologically marks the territory which he believes to be his own. From this we deduce the validity of interpreting the country house as a sign of possession (of and a focal point for) a reinterpretation of the subject in which the creation of artificial spaces set against natural ones is fundamental.

In an essay by Tomás Rueda Vargas about the Savannah of Bogotá there is a quotation from the French writer Gustave Flaubert which admirably summarizes his approach to (and feelings about) the subject: *There are corners of our land that we would like to embrace with our heart.* Rueda Vargas-born towards the end of the nineteenth century-reveals, in his prose, a remarkable sensitivity towards the political history, landscapes, peoples, usages and the fauna and flora of the countryside of the central Andean regions of the country, as do Armando Solano, Eduardo Caballero Calderón and Camilo Pardo Umaña. Others, like Manuel Ancízar, Vergara y Vergara, Jorge Isaacs and Tomás Carrasquilla also deal with country life in different Colombian regions.

In previous centuries, the colonial chroniclers created energetic geographical and statistical treatises or travellers' accounts about the countryside of Nueva Granada that are still fascinating. But most of them barely mentioned rural constructions and their relationship with the inhabitants of the countryside. It was the American and European travellers of the nineteenth century who first showed, in their books, a strong interest in country buildings.

The long-winded chronicler José Maria Cordovez, who extensively refers to the construction, appearance and use of the urban houses of Bogotá, barely makes a passing reference to the country estate houses of the Santa Fe high plains region or other parts of the coun-

Opposite page, *Antón Moreno* house set in landscape on the outskirts of Popayán, Cauca.

try. For every line which Tomás Rueda Vargas dedicated to hacienda houses in his chronicles, he wrote at least two hundred and fifty deliberating upon and describing thoroughbred horses and a hundred and seventy about more or less picturesque episodes of local political history, which inevitably take place in the melancholy atmosphere of high moorlands and prairies, with freezing but poetical rainstorms and sunrises.

In the majority of books written by Colombian or foreign authors, the presence of constructed forms is vague and sketchy, no matter whether they are dealing with humble thatched cottages or elegant houses in country estates. Always situated in the background, as settings or backdrops which apparently just happen to be there but always play a secondary role, country houses rarely appear for what they are in this literature, namely, as elements which strongly distinguish places and landscapes. On occasions, writers like Solano or Caballero Calderón refer, both with clarity and affection, to one country house in particular (either to *La Trinidad* or to *Tipacoque* in Boyacá, their native land), but these examples (which are cited in *Casa Colonial)* are, in the context of their literary production, exceptional. For them, and justifiably so, what is vital is the man in his house and in the countryside, not the country house of the man.

In their cultural formation, which is predominantly literary, the rural architecture of their own country, whether refined or anonymous, had little or no conceptual autonomy, which clearly predisposed them to assume, without a guilty conscience, an attitude of indifference towards or only a passing interest in the possible meanings or the inherent gracefulness of forms constructed in the countryside. The houses of the countryside were, for them, like the rain or sun, something that was there from the beginning, just another incident in the vast repertory of natural phenomena which made up country life.

This book is not intended to be a catalogue or inventory of country estate houses from the different epochs of Colombian or Spanish Colonial history. On the one hand, this task has been largely carried out, for many regions of Colombia, by specialists working on architectural inventory and analysis. And, on the other hand, the intention of the author of this essay is to provide both those with a professional interest in architecture and those who know little about it with a wider knowledge of a subject that is both attractive and only superficially understood.

While the vagueness or imprecision of the title, *Casa de Hacienda* (The Hacienda House), may bother purists of professional terminology, we hope that it will give the non-specialist reader a reasonably clear idea of what this book is about. Within the context of the series published by Villegas Editores (which began, a few years ago with *Casa Colombiana,* and later continued with *Casa Campesina* and, more recently, *Casa Colonial, Casa Republicana* and *Casa Moderna)* the subject matter of the present volume has a well-defined place, even though it may be a bit wide-ranging on occasions.

The subject of rural architecture covers much more than the houses of haciendas. There are also "farm" or "weekend" houses, studies of which will eventually be made and published separately, since there exist clear qualitative, dimensional, socio-economic and technological differences between the former and the latter. Beyond this, there are rural buildings which are not houses, such as mills, stables, religious residences, silos, storerooms, schools and even railway stations.

One part of this vast theme of rural architecture was dealt with in *Casa Campesina,* it being impossible to exactly define what is "popular or anonymous architecture" and what might be considered as "refined architecture". It would also be very difficult to make a precise distinction between the rural houses studied, for example, in the various books written by Alberto Saldarriaga and Lorenzo Fonseca and those which qualitatively surpass the simple condition of being "popular". The hacienda houses of the Spanish Colonial and Colombian periods constantly border on the two classes, showing technical and aesthetic characteristics that are common to both.

The hacienda has a socio-economic character that is different from that of the farm. In general terms, it shows different dimensions and aesthetics, which does not necessarily prevent many farms from having houses whose aspect and appearance allow them to be classified within the genre of hacienda houses. It is clear that the standing which the most notable hacienda houses have within the patrimonial inventory of the architectural culture of Colombia is superior to that which might be qualitatively assigned to the houses or cottages of farms or small-holdings, that is, *genuinely rustic* buildings, but this does not represent more than a convenient cultural convention. It would be fairer and more precise to give each sub-genre (haciendas, farms, small-holdings) a certain classificatory independence and autonomy, without employing arbitrary rankings such as the one which places certain "monuments" above the rest of the architectural culture of a country or an epoch. Or the no less absurd criterion which says that "cultured" or professional architecture is superior to that which presumably is neither one nor the other.

Several of the architectural works previously mentioned in *Casa Colombiana* and *Casa Moderna* are, in fact, rural houses and some of them also were, a long time ago, hacienda houses. We intend here to incorporate the republican period into the genre of country estate houses, leaving the rest of the subject for a later book because of its extense and complex character. The quantitative emphasis in buildings of the republican epoch of the country's history leans towards the urban, since the "new" hacienda houses built during the nineteenth century are less numerous than the transformations wrought upon already existing colonial buildings. Nor should we forget, for example, that the coffee-plantation house, an architectural development typical of the nineteenth century, is a quasi-autonomous genre.

The subject matter of this book was set forth in one of the chapters of the *Historia del Arte en Colombia,* and later, in the book *Crítica e Imagen* (1977). An enlargement of my criteria for hacienda houses of the colonial period in Nueva Granada was published in 1990, as part of the book entitled *Estudios sobre Arquitectura Iberoamericana*. The growth of the bibliography on the subject is now reflected, in a concrete way, in this more recent vision, which goes beyond the chronological limits of the colonial period.

There is now an enormous chronological and conceptual distance separating the author of this essay from an incident which took place in 1964, when his interest in colonial hacienda houses naïvely began. Upon asking, at that time, for more information about the subject, a history professor at the National University –an expert in "classical" (?) and neoclassical architecture– curtly replied: "What's the point?; this subject is not on the curriculum of the University". In fact, it was just one of several architectural subjects which went against the academic prejudices of that epoch. In that academic ambit the colonial architecture of the country seemed to be limited, at most, to a number of churches and cloisters and the fortifications of Cartagena.

The mistaken attitudes of that time were typified by an an inventory-type book, published in 1966 by the Coffee-Growers Cultural Fund as part of its "Colonial Heritage" series, entitled *Herencia Colonial en el campo colombiano,* which indiscriminately included a number of houses from the republican and modern periods, as well as a certain restaurant on the outskirts of Bogotá done in a farcical "neo-colonial" style. This was followed, however, by serious and meritorious professional studies, such as *La arquitectura de las casas de hacienda en en el valle del alto Cauca,* by Benjamín Barney and Francisco Ramírez, and *Guadalajara de Buga y su arquitectura,* by Jaime Salcedo, both of which we recommend to the reader who is interested in knowing more about the subject.

Within the general framework of the Colombian cultural heritage, rural architecture has, with only a few exceptions, always been in second place. The national panorama has unmistakenly been dominated by urban architecture. In the contemporary world-view, rural buildings tend to be regarded as working tools, "second houses" or something from "another world". If this study manages, even in a minor way, to correct this unfair situation and create a certain interest in the architecture of the countryside, it will have achieved the basic purpose that has motivated its publisher and all of those who have participated in its creation.

Previous pages, left and above, The house of *Coconuco,* Cauca. The hacienda, whose location resulted from an inspired choice of site, was consolidated in the first third of the seventeenth century.

The house of *Coconuco,* in its present form, dates from the middle of the eighteenth century. It has a compact volume, without complete perimetral galleries. As in the house of *Pisojé* and other haciendas of Cauca, a part of the single floor was given an upper level so as to supplement the owners' bedrooms. In functional terms, this allows for a certain visual command of the surrounding landscape.

Above, A beautiful environmental contribution from the republican period, a row of cypresses brought from France in the second half of the nineteenth century by General Tomás Cipriano de Mosquera borders the approach road.

Previous pages, opposite page and right above and below, *Casablanca,* Tocancipá, Cundinamarca. *Casablanca* is one of the most literal and repeated place-names in Neo-Granadian haciendas and is shared by eight houses in the high plains region of Cundinamarca and Boyacá. The house combined a floor plan organized around a central patio with exterior perimetral galleries. Note the similarity between the elevated section which forms a lookout balcony and those found in such houses in Cauca as *Coconuco, Yambitará* and *Pisojé*. The house is situated in a hollow in the landscape.

Left, As was the custom in the houses of Cundinamarca, the rooms originally lacked windows in the strict sense, having a single space for the window-door. One of these openings originally provided a visual relationship with the landscape.

Definitions, origins and precedents

HACIENDA. (From *facienda*). Agricultural property. 2. Accumulation of goods and wealth which one possesses. 3. Job, household task. 4. Work, action or event. 5. A matter or business involving several persons.
(Free translation, from the Dictionary of the Spanish Language, Spanish Royal Academy).

Although the Dictionary of the *Real Academia Española* mentions the various uses and meanings of the term "hacienda", there is no doubt that its most familiar meaning is the first, which is derived from the Vulgar Latin brought to the Iberian Peninsula by the Roman legions: *agricultural property.* The second meaning, which refers to accumulated wealth, brings to mind a socio-economic idea which is no less traditional. In the popular usage and colonial documents of Nueva Granada both meanings of "hacienda" were applied without distinction and they remained in force, without any modification, into the Colombian republican period. Even today large rural properties are called "haciendas" and the management of the economic resources of the Colombian State is, in theory at least, in the hands of the Ministry of *Hacienda,* an entity with an archaic name but modern inefficiency. The term "estancia", common in the south of the continent and in the early epoch of the colonization of the south of Brazil and Nueva España (Mexico), was never or only rarely used in Nueva Granada. Although the Andalusian term *dehesa* (which refers to a cattle, rather than an agricultural, estate) was employed during the colonial period in Nueva Granada, it fell into disuse in the republican period (19th century). In addition, the word "hacienda", incorporated into English as an hispanicism, was not unusual in those Spanish colonial territories which later became part of the United States (Texas, New Mexico, Nevada, Arizona, California).

The Mexican architect and researcher José Antonio Terán, in his *Arquitectura rural en México. Las haciendas de una región*[1], cites, from the "Dictionary of Authorities"[2] prepared for Nueva España in the eighteenth century, the following charming definition of the term "Hacienda": *Country properties and lands of labour that are worked to make them fruitful.* This colonial document thus confirms the idea, shared by present-day chroniclers and historians, of the hacienda understood as an autonomous productive unit which may or may not have buildings. The latter depends on whether the owner of the property habitually resides there or only makes occasional visits. In turn (according to the Spanish researcher Maricruz Aguilar García, in *Haciendas de Olivar,* appendix to the summer courses of the Antonio Machado University, Baeza, Spain, 1988), the buildings that may arise in the terrain of a hacienda belong to "*...a specific type of architecture linked to the natural landscape and a system of agricultural exploitation.*".

This precise concept allows us to link the socio-economic definition of the hacienda to the architecture which will eventually appear in it. A singular proof of this is the Anglo-Saxon term-borrowed from colloquial Spanish – that has been adopted in the Mid – and Far-West of the United States for large and small cattle farms: *Ranch,* from "rancho", which described (and is still used for) both the property and the buildings on it, without distinction, or only for the latter. The properties dedicated to agriculture in the colonies of North America received other names, according to their size and use: *plantation or farm.* The French terms which are approximately equivalent to the above allude to their own medieval origins, in that the idea of *domaine* (hacienda) refers to the socio-economic status of the feudal lord's domination of lands and servants and *ferme* (the equivalent of the English *farm)* or *clos,* that is, a closed precinct or area, designate what would be a plantation or farm.

In *Casa Colonial* the author of the present essay traced the historical origins of the hacienda of Nueva Granada to two disinct historical circumstances: *"One, the... system of ideological, technological and anthropological processes which make up... in the Middle East and around the Mediterranean, the spatial archetypes of dwellings that would eventually cross the sea. And the other, corresponding to the specific circumstances surrounding the implanting, on this continent, of a certain version of the architecture... existing at the end of the fifteenth century in the centre and south of the*

Opposite page,
Farm on the outskirts of *Olvera,* province of Cádiz, Andalusia.
"It is not Africa which begins in the Pyrenees. It is America which begins in La Mancha".

Iberian Peninsula ... The 'what' of the house (of a hacienda) is explained by tracing its distant architectural ancestor and the 'how' is explained by surveying the specific process through which Spanish houses were brought to the New World".

Twentieth century archaeology has been discovering more and more traces of the buildings of the most ancient cultures known in the Middle East and around the Mediterranean, which necessarily had to be rural ones, at least in their beginnings. Thus, in a very indirect and generic way, we might (stretching things a bit) consider a presumed Sumerian or Babylonian "country house" as an historical precedent for the Latin American hacienda house, given that anthropology confirms that the relation between man and the countryside has changed very little in the last thirty centuries. We may add, to this archaeological reflection, the fact that there exist, in the recently-uncovered pre-Columbian urban centres of Teotihuacán, in the valley of Mexico, and Chan Chan, in Peru, clear signs of a systematic division of rural areas relatively far from the urban centres, some showing traces of buildings. Different conclusions may be drawn from these very interesting discoveries, including a strong possibility that the equivalent of country estates or farms existed in the pre-Hispanic epoch, at least in the valley of Mexico and in Upper Peru.

It is worth recalling that the socio-economic relationship between man-countryside-city has taken two basic forms throughout history, apart from the anomalous circumstance of the absentee landlord: the first, in which the owner or user of rural properties permanently lives and works on them and the second, in which he permanently lives in some urban nucleus more or less near the property and is only in the countryside on a temporary basis.

The other historical factor, mentioned above, has to do with the process whereby rural architecture was exported from metropolitan Spain to its imperial possessions in the New World. The process of Hispanic imperial conquest and colonization was necessarily preceded by the formation, in Iberian territory – as the researcher Maricruz Aguilar indicates – of an architectural genre characteristic of the landscapes of the peninsula and suitable to the specific activities of exploiting the land. As would occur, centuries later, in the New World, the rural architecture of Spain did not arise through spontaneous generation nor as the result of an endogenous process. It was begun, in a limited way, by the Greek and Phoenician settlers who established themselves on the coasts of what are today Andalusia, Levant and Catalonia, although the systematic possession of uninhabited areas and the beginning of constructed nuclei that were capable of a self-sufficient agricultural production were really the work of Roman colonizers. This is evident when we consider that such a process was part of the simultaneous creation of a system of territorial dominion and exploitation that included land and river communications, as well as large-scale hydraulic and public works.

In short, the Roman colonial empire laid the socio-economic and political bases for the historical process which imperial Spain would have to repeat in America sixteen centuries later. In Hispanic territories socio-economic circumstances and tendencies arose which allowed for the remaking of what in metropolitan Rome was called the *latifundio* (a combination of two words from Vulgar Latin, latus, broad or ample, and *fundus,* a rustic farm), that is, a rural property of great size. Lesser properties were also established, namely, the "hacienda", and the smallest in terms of size, the "finca" or farm. What came to be refined during the centuries that followed were not the basic principles of political dominion over the land, which, by their very nature, could not be changed very much, but the modalities of property, its management and systems of exploitation and production, thus making up the historical process which Spain would be destined to repeat in the New World.

The Roman use of rural areas (whether metropolitan, provincial or colonial) included adapting to the countryside dwellings that had originally been created for cities, towns and villages, since it was not difficult to transfer, from one medium to the other, schemes of spatial ordering that were highly versatile in themselves. The generic term "Villa" came to be used popularly and without distinction for both a modest weekend house in the country or an enormous palace-like ensemble built on the outskirts of a city, as was the case with the "Villa" of the Emperor Hadrian near Rome or the famous "Villa" Armerina. It is obvious that the managing of rural properties brought with it the need for constructing buildings for dwelling and work different from those which were used in holiday retreats, so that the popular term "Villa" was extensively used for utilitarian rural structures. The Spanish historian Fernando Chueca Goitia mentions, in his *Historia de la arquitectura española*[3], some archaeological discoveries of rural "villas" that date back to the middle of the second century A.D.

He remarks that: *"As well as aristocratic villas there existed in Spain (Hispania) farms or rustic villas which are known through the writings of Varro and Columella. In general, all of the buildings surrounded a courtyard; near the entrance were found the rooms of the 'villicus', the official who looked after the estate on behalf of the landlord. There were rooms for servants and slaves, a kitchen, bathrooms and a jail, as well as the storerooms and outbuildings necessitated by agricultural labours (for grains, wine, olives). Remains of these rustic villas are conserved in Catalonia, where agricultural exploitation flourished..."*

The above quotation, without any modification, might also be employed to describe the house of a Mexican or Neo-Granadian hacienda, not in the second century but in the seventeenth. The Roman colonizers, having learned all this from the Greek settlers of the eastern Mediterranean, did not do anything other than rework the ancient spatial ordering traditionally practiced by the oldest cultures of the Middle East: the ground floor of the house organized around a free central space (the patio) and the house with a concentrated floor plan, in which one or two wings of rooms were surrounded, on their perimeters, by colonnades or peristyles.

The historian Chueca Goitia adds[4]: *"The most complete example of a countryside (that is, rural) villa is that of* Cuevas de Soria. *The rooms, many with an apse-like plan, are arranged around an enormous peristyle of Tuscan columns. The large reception hall, the "triclinium"* [banquet hall] *and steam baths stand out...It was probably built in the middle of the second century and abandoned in the middle of the fourth century..."* The above could equally describe one of the larger Mexican haciendas of the eighteenth century, including the use of Tuscan columns and the separated location of the enormous dining room and bathrooms.

The historian Chueca Goitia continues: *"In the third century we find a very remarkable example of a country villa in Almenara de Adaja, province of Valladolid* (an Arab name, used, centuries later, for the place where the remains of the Roman house referred to were found). *Its floor plan reveals an architectural conception (of) ...spaces that are intertwined with a subtle mastery."*. The above implies that, despite being scarce, the archaeological vestiges of rural buildings found in Spain are sufficient to establish a qualitative architectural hierarchy that was to be repeated, centuries later, in the New World. The "Big Houses" of the Brazilian *fazendas* and the enormous palatial residences of a number of Mexican haciendas, together with the more humble country houses of Nueva Granada, have very ancient Iberian ancestors that are perfectly identifiable as such and display their respective categories of importance.

The country houses of the Roman colonies were obviously used by the Islamic nations which spread from the Middle East towards the centre and south of what had been the colonial province of Hispania. The Islamic cultures were more inclined to conserve and modify the architecture of the territories they invaded than the Vandalic ones, which destroyed the buildings, not only of the conquered peoples but also of their ancestors. In much the same way, the ruins of the colonnaded halls of Persian palaces are probably the source of the spatial ordering, rigorously compartmented through structural modules, used in the mosques built around the Mediterranean, from Mesopotamia to Cordova, in Hispania. The successive Islamic invasions of Iberian territory between the eighth and fifteenth centuries and the establishment of a settled Islamic culture and urbanism also gave rise to a long process of adaptation and management of the vast rural territory of what was then called Al-Andalus.

Few architectural genres turned out to be more suitable to (and in harmony with) the sense of existential order and the customs and forms of exploiting the rural territory of the Islamic peoples than the country house, whether recreational or workmanlike. For nearly eight centuries, the Iberian land witnessed a flourishing of agriculture, supported by the use of hydraulic works that revived the technological traditions of Roman engineering, to which one may add Arab arts and sciences, which were ignored and even prohibited by the religious obscurantism of the Christian nations of Europe. The extraordinary environmental nature of the Andalusian countryside in the epoch of Islamic domination gave rise to the famous legend - which might have been perfectly true - that a squirrel could go from Cordova to Granada *without touching the ground,* that is, from branch to branch through interminable and marvellous forests. Such an ecological management and care of territory, taken by nations which were originally nomadic desert tribes and "infidels" to boot, is extremely thought-provoking. The great contribution to history of the rural civilizations of the Islamic peoples of Spain was that they turned the countryside into an emporium of wealth, while simultaneously conserving the beauty and ecological equilibrium of the landscape. It is no coincidence that up to the present day, when one speaks of water in the countryside in Latin America, charming Arabic terms that transcend the centuries are still employed, such as "acequias" (irrigation trenches), "atarjeas" (culverts), "alcantarillas" (sewers) and "aljibes" (cisterns).

The Roman country houses were largely vandalized and destroyed by the Visigoths and the early Christians in Hispanic lands, in accordance with what might be considered as their "normal" habit of obliterating what they found. It was more logical for them to dominate and defend rural territories by means castles and fortified towers. In short, after the the dissolution of the colonial Roman empire there arose a five hundred year-long territorial domination, by the Visigoths, of the centre and south of the ancient Roman colonial province of Iberia, one that was interrupted by the Islamic invasions. This led to a notable gap in the accounts and analyses of the historians who have devoted themselves to the subject of civil architecture in Hispanic lands. Spanish specialists like Torres Balbás and Chueca Goitia[5] have been forced to directly relate the few vestiges of rural architecture of the Roman period to those of the Islamic farmhouses, ignoring, for obvious reasons, the historical void represented by the Visigoth and Paleo-Christian epoch. It is worth noting how the historiography of this epoch concentrates, not on urbanism or urban or rural dwellings but rather on religious and military architecture.

The above does not at all mean that the process governing the appearance of the regional Iberian forms of *rustic* building did not occur in the period or was interrupted by these historical and political events. On the contrary, the formal structure of "popular" or regional Hispanic architecture began with (and continued, without interruption, after) the dissolution of the colonial Roman regime. The primitive elemental and temporary nature of the country house, which has its origins akin to a refuge following some shipwreck, has permitted it to arise and survive despite the upsets of history and in the most unfavourable circumstances imaginable. The great advance to be noted is that which separates the lean-to or the temporary shed from the permanent and definitive house, since they are two highly different rural architectural genres. The hacienda house is not, qualitatively speaking, an upgraded cottage but an architectural species created to replace the former.

The "alquería" (from the Arab al-*karía,* cultivatable land, according to the Dictionary of the Spanish Royal Academy) is none other than the Roman "villa", used, without distinction, for a vacation house or a productive country unit, or even as a mixture of the two uses. This is logical, when we consider that, in the Muslim kingdoms in Hispanic lands, the ruling classes governed a socio-economic system that had at least as many similarities to, as divergences from, the feudal forms characteristic of the Christian kingdoms. While it was necessary, for reasons of basic collective security,

Opposite page, House of an olive-plantation hacienda and an olive-oil mill of the eighteenth century on the outskirts of Lupión y Bejigar, province of Jaen. Patio between the manor house and the permanent workers' quarters.

to create walled precincts and fortified citadels and the "alquerías" frequently had protective walls around them as well (as occurred, centuries afterwards, in the haciendas of the north of Mexico), this did not stop them from developing into a system which managed the cultivation of grains, grapes and olives and exploited (in a selective and moderate way) timber forests and hunting and fishing reserves. It was the Islamic nations which gave a definite shape, in Hispanic rural history, to the landlords' custom of double residence (and, as a consequence, of double life): the city and the country house. The extraordinary culmination of this process was the decision, by the Caliphs of Cordova, to build what would be the Andalusian branch of the earthly paradise, the enormous rural palace of Medina-az-zahara, situated at nearly two leagues from their seat of government in the city.

The social relationship of man-countryside that was eventually brought to the New World by the Spanish colonizers reaches its high point in Andalusian territory with the Islamic cultures. Rome had laid down the basic principles for the exploitation of rural lands, which were later exported to Hispania, but they had to be modified and refined by the Islamic cultures so that the Hispanic empire could send them, in turn, to its possessions in the New World as a socio-political synthesis within which the country house was only one of the tools needed to strengthen the continuity of what was already a long tradition.

The prolonged period of wars and territorial and social instability which marked the end of the medieval epoch and of the Islamic kingdoms in Hispanic territories did not favour the survival of country houses or the creation of new ones, so that only a few managed to weather such difficult historical conditions. Indirectly, this may be the fundamental historical reason why it became necessary to reinvent or (if you will) resurrect the hacienda, the farm and the ranch in southern Spain towards the end of the fifteenth century, once the wars of reconquest of the territory formerly dominated by the Moors had ended. It would have been natural for the Castilians, Aragonese, Galicians and Extremadurans to first reject the forms of dominion over (and control of) the countryside practiced by the Moors and Saracens and then, in the absence of other models, accept the validity of (and reinstall part of) what had existed before and adapt it, cleverly or crudely, to their own idiosyncrasies and limitations. The Andalusian historian Antonio M. Bernal[6.] thus supports the historical hypothesis that the appearance of the *cortijo* was an agrarian phenomenon that followed the wars of reconquest in Andalusia, that is, one that dates from the beginning of the sixteenth century. He also supports the definition of the *Cortijo* as a unit of exploitation of a raw material, such as cattle-farming, in which there does not exist a process of transformation like that which is found in haciendas that grow olives or grains. This hypothesis rejects the idea that the Andalusian *Cortijo* was a precedent or forerunner of what might have been its equivalents in Nueva Granada, since the two developments were contemporary. It is worth noting that the term *Cortijo* was never, or only rarely, employed in Nueva Granada: it was only revived towards the end of the twentieth century as an "hispanified" affectation among owners of Colombian cattle farms.

It is not surprising that, later on, in the Bourbon epoch of the eighteenth century, the Spanish monarchy would see the need to undertake colonization of some of its own metropolitan territories, which had been left to their own fate for a long time. The nation which had managed to create an empire on which the sun never set had completely forgotten about regions such as those which ran along a good part of the road between Madrid and Seville or were situated in Jaén, where it had to do, in a hurry, what had already been achieved two hundred and fifty years before in the New World: establish cities and towns, methodically distribute rural lands, create haciendas, protect water sources and forests, cultivate and raise something more than olives and fighting bulls and even try to introduce timid agrarian reforms in order to improve the permanently critical state of the Spanish countryside.

The abandonment of the hydraulic systems and works of the Roman and Arab epochs led, as a long-term historical consequence, to the socio-economic backwardness, in comparison to other parts of Europe, of the rural parts of central and southern Spain. It was only in the nineteen-forties that a remarkable and superhuman effort was made to supply the country with reservoirs, dams, irrigation canals and water reserves needed to revive agricultural production and give new life to lands which, over the centuries, had gradually turned into a branch of the Sahara desert. In half a century Spain had to correct a situation that had worsened over half a millennium, a good part of the latter period coinciding, paradoxically, with the height of its imperial power.

On the other side of the ocean, in what were called the "overseas provinces", from Nueva España (Mexico) to the jungles of Paraguay, it was the Jesuits, not the lay settlers or colonial administration, who revived the agricultural techniques and knowledge of the Islamic "infidels" and applied them to the colonies of the New World in order to optimize yields and maintain a reasonable economic equilibrium in their haciendas. This mastery of agriculture was seen in the conservation of vegetation around water sources, a correct management of hydraulic resources, the preservation of forests, crop rotations, the alternation of fallow and pasture lands and the rational use of available manpower, all of which allowed the Jesuits to develop and exploit farms, estates and enormous *latifundios* with an efficiency and economic benefit that would eventually arouse the envy and ill will of the holders of land grants, estate owners and colonial governments and was one of the basic reasons for their expulsion from the overseas provinces.

The vestiges of Islamic rural architecture in Spain are even more scarce than those of the Roman epoch, given that the Andalusian "alquerías", never very numerous, became victims of the wars of reconquest and the vindictive and vandal-like feelings of their Christian enemies. More abundant were the towers or "alcázares", built by Christians or Muslims, that were used as recreational houses in the countryside and, at the same time, served as a fortified refuge, although these were not immune to the destructive spirit of recent epochs either. Chueca Goitia[7] (op. cit.) mentions an "alquería", or to be more precise, a general-purpose "hacienda" or "almunia", that is, a non-fortified farm, in which a holiday house was combined with a working farm. This writer says: "*Further away from the city (Granada), on the Zubia road, exists a beautiful almunia, which was known as 'Darabenaz' in the Muslim period. It is made up of two wings that form an angle and create a courtyard that looks out on the beautiful mountain landscape...*". This description exactly serves, as well, for the "alquerías" of Nueva Granada, such as the estate houses of *Calibío* or *Antón Moreno,* on the outskirts of Popayán, or *Cañasgordas,* near Cali.

Even the medieval genre of the house-fort was influenced by the patterns of spatial ordering which were found in the contributions of the successive cultures that had colonized the Hispanic countryside. Throughout Spain, there are numerous examples of ruined or reconstructed "castles" whose interior spaces are arranged around a central rectangular courtyard and in whose perimetral wings the great architectural contribution of the Muslims made its appearance: the "alcoba" (al-quba), that is, the Islamic version of the private Roman space: the bedrooms or *"cubiculae"*. This breaking-up and definition of the spaces which, in preceding medieval buildings that were strictly military or defensive in character, had been generic or undifferentiated is of vital importance to the history of both urban and rural domestic architecture.

At the moment, it is possible to find the origins of the Neo-Granadian hacienda house in archaeological vestiges and historical documents, but the search for its direct architectural precedents in Spanish territory is necessarily sterile. Some Latin American researchers have pointed out – and, with a certain naïvety, lamented– the *current* absence of typologies or rural constructed forms that would allow for a philological and chronological linking of what might be Spanish precedents and later versions of the same in the New World. There are three basic reasons why this is not possible. The first is that any attempt to apply the formalistic and chronological analytical system developed for religious, institutional or military architecture to a "popular"

architectural genre that is formally and technologically non-evolutionary and non-monumental is destined to failure by definition and\or because of its conceptual basis. The second, which derives from the first, is, simply, that the Andalusian olive-grove hacienda, the vineyard hacienda of La Mancha or the *Cortijo* found in parts of southern Spain are architectural species that are very different from the sugar-cane hacienda of the Cauca Valley or the grain-growing one of Boyacá: they play different roles in their respective societies and economic systems.

A number of Spanish researchers have pointed out how the Andalusian olive farms or mills of the eighteenth century turned into veritable factories for producing oil that was largely exported to the colonial domains of the Spanish Crown, where the growing of olives was strictly prohibited in order to protect the agricultural economy of the metropolitan region, which was excessively centred upon the production of olive oil. This observation allows us to deduce that there were enormous functional and formal differences between the Andalusian centres of production and the country houses of Nueva Granada, where this product was an expensive article of consumption. The large Andalusian country house-factory, as it arose centuries ago and is still observable today, did (and does not) have a Neo-Granadian colonial equivalent. As far as rural architecture goes, the history of Andalusia during the past four centuries does not show any concrete resemblances to that of the province of Nueva Granada. After the conquest of Granada and the "discovery" of the New World, events in the Andalusian countryside and those in the overseas provinces followed divergent socio-economic courses. The Spain which set out to conquer a hitherto unknown world was one thing, while the Spain which, in highly diverse ways, chose (or resigned itself) to stay linked to its European destiny was quite another.

What we see today in the centre and south of Spain is the result of the above. In formal and technological terms, the hacienda houses of Nueva Granada show spatial and constructional (and by extension, aesthetic) characteristics that are difficult to find in their rural counterparts in Andalusia, La Mancha or Extremadura. On the other hand, such parallels are abundant in *urban* houses, where it possible to find the chronological precedents which one desires. Practically all of the constructional systems, materials and finishings seen in the hacienda houses of Nueva Granada can be found in village and city houses in Andalusia or Extremadura. Indeed, it would be strange if this were not so. We shouldn't ignore, of course, the extreme interchangeability (between the urban and rural) of the constructional systems and spatial ordering seen in Neo-Granadian architecture, although this does not mean that the hacienda house partly derives from the city one, but that both have a basic common ancestor.

The third reason that can be invoked in this case is chronological: the Colombian historian Germán Colmenares, in studies such as *Cali: Terratenientes, mineros y comerciantes* and *Popayán, una sociedad esclavista*[8], correctly points to the second half of the seventeenth century and the first half of the eighteenth as the period in which the "definitive" versions of the colonial hacienda houses of the two regions arose or reached their highest development, replacing cottages or huts that had been erected, in some cases, as temporary shelters and in others, as new buildings in the territory. In turn, Spanish historians like Antonio Bonet Correa or Maricruz Aguilar state that the great epoch of the Andalusian haciendas and *cortijos* evidently occurred in the same period noted for Nueva Granada by Germán Colmenares and the researchers Benjamín Barney and Francisco Ramírez[9]. In short, the haciendas, farms and *cortijos* of the south of Spain are, in their majority, contemporaries of their Neo-Granadian equivalents, which obviously rules out the possibility of their being forerunners. The superposition or original building of baroque decorative elements in the Andalusian hacienda houses, in many of the cases cited by Antonio Bonet Correa in *Andalucía Barroca*[10], are *later* than the building of some of the haciendas of Boyacá and the Valley of Cauca illustrated in this book. It should be noted that, in the case of certain enormous haciendas in parts of Mexico, there arose an addition of baroque decorative elements in local versions that was

Opposite page, The original enclosure walls of the complementary spaces in *Canoas,* Soacha, Capital District.

analogous to what happened in the same period in Andalusia. In Nueva Granada, where the immense wealth and rural development of Nueva España never came into existence, such stylistic phenomena could never have taken place.

For unknown reasons, historians and students place little importance upon the individual and collective memory of the Spanish colonizers, an aspect of the origins of rural colonial architecture which is decisive and vital for the genre of rural architecture. The many-sided memories of the settlers of the countryside of Nueva Granada must have led them into obsessive analogies of landscape in their interminable search, sometimes happy and sometimes desperate, for places where they might find, in the vast geography of Nueva Granada, lands which the European continent had denied them and suitable sites for the foundation of cities.

In the case of ambitions for rural lands there is no other way to explain the remarkable analogy, noted in *Casa Colonial*, between the nearly identical panoramas around Medina-Sidonia (in the province of Cadiz) and Villa de Leyva, in Boyacá. There is no end to such examples: the "desert" of La Candelaria, in Ráquira, Boyacá and the region of Tabernas, in Almería; the sierras of Ronda and those of the north of Boyacá and the south of Santander; the region of Río Tinto, in central Andalusia and that of Piedecuesta and El Chicamocha in the south of Santander; the countryside around Guadalupe and Oropesa (province of Toledo) and that of the valleys of Sogamoso in Boyacá or Pubenza, around Popayán; the prairies of Extremadura and those of Tolima and Huila; the Valencian plain and the valley of the River Cauca; the Sierra Nevada of Granada and that of Santa Marta, on the edge of the Caribbean Sea ... The coincidences are too notable and numerous to be a mere matter of chance. In *Casa Colonial,* we read: "*... One may imagine the strong and heartfelt emotions of the tough Spanish settlers when they saw places which seemed to have been miraculously transported over the Ocean sea from their native lands...*". But there, in what was thought to be "The Indies", where the land seemed to have no limits, the annual seasons, regulators of European life and customs - including the relation between man and the countryside and even the life of seeds - did not exist. The frozen high moorlands, the high plains of eternal spring, the tropics everlastingly reminiscent of the fierce Extremaduran summers gradually replaced their memories of Europe. The physical appearance of the Neo-Granadian countryside might have brought memories of Iberian landscapes, but it had a separate existence, as a New World, or, rather, as something out of Another World. It could be, at one and the same time, paradise and hell, with its fertile lavishness and terrible demands.

The Hispanic memory obviously included the country house as a means for taking possession of, dominating and exploiting the land. The underlying wish had to be that of dominating the new lands which had been seized in line with European will and practice, creating in the neo-Hispanic countryside islands of life brought from the other side of the world, just as the first towns of Spaniards were seen as Andalusian or Castilian urban microcosms encrusted in the most unlikely parts of a geographical medium that seemed to belong to another planet. A vital part of that memory, partially conscious but intuitive, would also dictate the least explicable and most poetical characteristic of the house in the midst of the countryside: its sense of place, that is, its complex physical, environmental and sensory relations with the landscape around it. The patterns which, in Andalusia, La Mancha or Castile, governed the relation of house-landscape, or natural space-physical space, were recreated in the New World on the basis of memories of how things had been in the Spanish countryside, so that while the horizons and dimensions of the natural space might have varied, the role played by the buildings erected in the landscape stayed the same. The further away the Spanish settlers were from towns and cities and the stronger the solitude of the countryside, the more powerful and influential became the presence of memory. We only have to look again at the vital enclosure of walls around the hacienda houses of Nueva Granada to be reminded of the origins of those walls: according to Fernando Chueca Goitia[11] (op. cit.) "*The problem for an Arab is not providing himself with a roof but an adobe wall. He can live without a roof but he needs walls. Once he is inside his enclosure, which is the truly indispensable thing, his requirements for comfort are minimal..."* And if we apply it to the general nature of the Neo-Granadian country house, the rest of the above quotation is fully valid: *"This special attitude towards comfort is also found, in a certain way, in Andalusian life. The Andalusian house manages to delight the senses, principally the vision, but it ignores the most elemental principles of functional comfort".*

The Neo-Granadian country estate house will thus be a working tool, a gesture of defiance and possession and a tight web of ambition, magic and illusion, as well as the physical form which gives validity to memory, supports customs and serves as the setting for familiar events. It will also have a mysterious quality which distinguishes it. In it one feels presences - and absences - which science now pretentiously calls paranormal ones, when, in fact, they are what lend a bit of interest and excitement to the prosaic normality of everyday existence. Any ancient hacienda house deserving of its name must possess certain dark and mysterious corners, occult and forgotten stories and a ghostly presence or two, which might be that of the cathedral canon Ignacio María ("Nacho") de Tordesillas y Fernández de Insinillas, on whose initiative the house of *Fusca* (Torca, Cundinamarca) was constructed. Being a documented and photographed ghost, the canon is a more important personage in the history of the house than the Liberator Simón Bolívar, who briefly lived there. Life did not allow the Liberator to build a beautiful hacienda house: he always preferred those which already existed. The ghosts might also be the night riders described by Armando Solano in the house of *La Trinidad* or the more or less evil spirits which have occasioned repeated and futile exorcisms in rural houses throughout Nueva Granada. And what would an old house in the savannah of Bogotá be like without the plaintive cry, just before dawn, of the ghost of some unknown murder victim that rises from an unsanctified grave?

The country house is born from the remembrance or family tradition which establishes how a house must or can be, and where it should exist and why. It also requires memory and a technological tradition in order to know how to build it and memory or a stubborn family tradition so that it may be cared for, loved and understood. This is, perhaps, the most profound and valid origin of the country estate house, though it is one which is ignored in books on the history of architecture, since accounts of how constructed forms came into being rarely take human nature and conduct into account.

...a specific type of architecture linked to the landscape and a system of agricultural exploitation. Maricruz Aguilar.

Left and above, *Los Laureles.* Facatativá, Cundinamarca. A house on the Savannah of Bogotá. whose environmental grace exemplifies the definitions and origins of the Neo-Granadian hacienda house. The laurel was one of the European plant species brought to the New World by the Spanish settlers who later founded haciendas. The inspired builder of the house deserves being called a "poet laureate" because of the splendid sense of place it displays.

Above and right, *Los Laureles.* Facatativá, Cundinamarca. It is an "up and down" house built around a patio. The gradual growth of the house is evident from the informality of the relationships among the volumes which shape it. The upper galleries around the patio were "republicanized" and now display flat ceilings that are inappropriate for houses of the colonial period. Letting the vegetation dominate the constructed forms may be an indirect Islamic influence.

Left and above,
The house of *Cuprecia,* Santander de Quilichao, Cauca. Although considerably modified in recent epochs, *Cuprecia* is one of the few surviving examples of a Cauca "low house". Its construction only included three sides of the central patio. The galleries around this and in the exterior façades show a surprising spaciousness, being more important than the spaces of the bedrooms and halls.

Opposite page, above and below, and right, The house of *Guacarí,* Cauca Valley. Strictly speaking, it is no longer a rural building, nor does what exist correspond to the first seat of the hacienda of that name. It is possible that some of the architectural components of the original house have been incorporated into the present one, which was erected when the hamlet of Guacarí grew up around the hacienda built in the seventeenth century and it became necessary to construct a church there. At the beginning of the nineteenth century the building became the local priest's house. The free surrounding spaces, which are unusual for an urban setting, are what one would expect of the house of a farm or a hacienda. While the enchanced volume of the upper storey is characteristic of the rural houses of the region, the recent restoration of the house exposed the untypical depressed brick arches of the galleries of the ground floor, leaving the house with an equivocal decorative accent.

Opposite page and right, *Baza,* Tenza Valley, Boyacá.
A house which "lets live" the countryside around it, in the manner of the Arab-derived "alquerías" (farmhouses) of Andalusia. Note, below on the left, the variation, usual in Boyacá, of a technically primitive roof frame built with the "par y nudillo" technique (see note No. 30), done with round rafters that are exceedingly slender as "pares". These rafters, which are very flexible but light and cheap, bend beneath the weight of the tiled roof and require braces or struts (that is, even more round rafters) resting on the tie beams to hold them up. Such a confused artisanal system reflects the picturesque nature of technical ignorance and the amusing side of improvisation. To the right, the access to the patio is a clear architectural success which reflects the art of knowing how to provide an entrance.

Left above and below, and opposite page, *Yambitará,* Popayán, Cauca.
The house is now practically surrounded by the city, though it originally lay at half a league's distance from the urban centre. With an L-shaped ground plan, it has an elevated section typical of hacienda houses in Cauca. It also has an interesting aqueduct that feeds the open-air bath (left) made of stone and brick, as well as generous perimetral galleries. *Yambitará* conserves a small part of the countryside which originally surrounded it, including the hill on which it stands, although it is now inevitably close to the housing developments resulting from the growth of Popayán.

Opposite page and above, the house of *San Pedro Alejandrino,* Santa Marta, Magdalena.

For reasons that have little to do with architecture, the house of San Pedro Alejandrino is unique in Colombia. The extensive sugar-mill hacienda and its modest house had reached its maximum point of development several decades before the Liberator of Colombia, Simón Bolívar, died there in 1830. Shortly afterwards, the dismembering and progressive destruction of the hacienda's seat began, the only part of it that was reasonably preserved being that section which contained the bedroom where Bolívar died. Towards the end of the nineteenth century a progressive monumentalization of the site began, starting with the surroundings of the house itself. Later on an "Altar of the Fatherland" arose there, then a gigantic parade ground and, finally, an implausible museum of modern art, all having little to do with the colonial hacienda house. The presence of Bolívar saved the house from disappearing, but, to honour him, the new Colombian republic created a monument around it which scandalously jars with the austere setting that witnessed the final hours of the Liberator and gave San Pedro Alejandrino a significance which has little or nothing to do with its original architectural traits and merits. The well-kept gardens and splendid tropical vegetation that make up the sort of elegant goldfish bowl that now surrounds the house of the colonial landlords of San Pedro belongs to a world that is very different from the environmental informality which it must have had in the eighteenth century.

The historical roots of the house of "los señores" of the chocolate-bean and sugar-cane mill hacienda of San Pedro Alejandrino are revealed by its flat (or terraced) roofs, which originated in the coastal towns and cities of the south of Andalusia and Levant in Spain. The mill, workers' or slaves' quarters and granaries are different, being roofed with wooden frameworks that show variations of the Islamic system of "par e hilera" (see footnote No. 30).

Opposite page, The house of *San Pedro Alejandrino,* Santa Marta, Magdalena. Note the strictly utilitarian architecture, devoid of decorative accents, in one of the perimetral galleries…

Right, Interior of the landlords' house, with furnishings that presumably correspond (or are similar) to those of the epoch in which the Liberator Simón Bolívar died there.

Hacienda and house

What we commonly call nature is not natural, it is the product of human activity, that is, artificial. According to Emilio Sereni: *"The rural landscape is the form which man consciously stamps on the natural landscape by means of his agricultural activities."*
GISELA VON WOBESER.[11A.]

The historical structure of the haciendas and, therefore, of the houses situated in them, is not independent of the process by which the province of Nueva Granada (which covered the present territory of Colombia and part of Venezuela and Ecuador) received its socio-economic formation. The special characteristics of the general history of Nueva Granada are what indirectly establish the differences between the rural buildings of this province and those of other regions of the Latin American continent. Country houses are invariably the image and likeness of the human group which builds and uses them, reflecting individual and collective strengths and weaknesses. In Nueva Granada the hacienda was one of the bases of the colonial economic system, although it never achieved the development and influence characteristic of the large Mexican or Brazilian *latifundios.* On a regional level, in an economy restricted above all to agriculture or cattle-raising and without the enormous mining production of Nueva España or Alto Perú, the Neo-Granadian hacienda was a socio-economic factor of prime importance, given the absence of industries and industrial production, whose existence were unlikely in the geographical ambit of the province. Urban centres, of which there were many in Nueva Granada, were, during the Spanish colonial period, points of consumption for rural agricultural production, but the transformation and external marketing of possible surpluses of such products were almost non-existent; colonial industry in Nueva Granada mostly consisted of crafts activities. The Neo-Granadian towns and cities lived off the countryside but gave little back to it.

The conquest of the territory of Nueva Granada and its later domination and colonization were chaotic processes, whose history is as confused as it is interesting, but a detailed description of such complex socio-political phenomena goes beyond the limits and focus of this essay. All we can do here is underline, as a general historical factor, the ambitious anarchy of the Spanish *conquistadores* as they sought to take possession of all the lands within their reach, driven by an exaggerated and boundless lust for wealth and material possession. This was only checked and subjected to a theoretical control by the Imperial Crown belatedly, when armed robbery and the law of the jungle had been in effect for several decades and the figure of the *conquistador* gave way to that of the colonizer and the holder of an "encomienda", a grant of land with indentured native labour.

The first metropolitan attempt to put the conquest of the new territories into some kind of order, the so-called Laws of Burgos, were proclaimed in 1512, that is, twenty years after the first "discovery" by Admiral Christopher Columbus. But it was only in 1573 that King Felipe (or Philip II) of Spain signed the version of the so-called Laws of the Indies in which, for the first time, a rough draft of official rules about the possession and hereditary management of rural lands was laid down. By this time, in the countryside of Nueva Granada, more than forty years of anarchy had passed, characterized by expeditions, punitive or possessive invasions and arbitary sharings-out of land, more or less done at random or by the force of arms. All this in addition to fraudulent or deceitful scrambles and manoeuvres as well as numberless legal tricks whose purpose was to consolidate and justify the possession of lands by any means. To say nothing of the subtle or crude dodging of the will (or improvised regulations) of the colonial authorities. The Colombian historian Margarita González, in her *El Resguardo en el Nuevo Reino de Granada*[12.], points out that it was only in 1680, with the compilation of the Laws of the Indies, that there came into being, at least theoretically, a broad, systematic and permanent legislative code governing rural possessions and goods in the overseas provinces. By that time the process of forming and developing the Neo-Granadian haciendas had been going on, more or less independently and without supervision, for about one hundred and sixty

Opposite page, *La Iraca,* Tibasosa, Boyacá.

years, which leads one to suppose that it was plagued with difficulties. The absence of a chronological and functional coordination between governmental and administrative theory and colonial reality is one of the most characteristic features of the history of the Spanish empire. The peoples of Castile, Extremadura or Andalusia knew how to grab hold of what initially seemed to be an enormous tract of the earthly paradise and their enterprise did not lack bravery and steadfastness, but the administering and governing of what soon came to resemble a purgatory and later a hell was a challenge which the Spanish empire failed at by a wide margin.

In the countryside of Nueva Granada, there was an inevitable territorial conflict between what in the Spanish colony were called "Los Naturales" (natives), that is, the indigenous groups, and the invaders – the European conquerors and (later) colonizers. Priests (whether missionaries or not) and some colonial authorities were – or at least pretended to be – concerned about preserving the indigenous cultures and their lands in various regions of Nueva Granada, in line with the Royal mandates to this effect, which were mostly the result of the writings and political pressures of people like Padre Bartolomé de Las Casas (1474-1568), a missionary and historian who worked in Mexico and called attention to the plight of the Indians.

But the reality was that the waves of Spanish invaders in the new territories had no reason to regard the indigenous population as anything other than the only potential source of manpower available for exploiting lands which, in its absence, had little value or productive capacity. What is more, indigenous manpower in Nueva Granada turned out to be much scarcer than in other parts of the continent. It was rapidly decimated because of the excessive workloads and ill treatment to which the natives were subjected and also due to the effect of the diseases which the Europeans brought to the New World. This is the reason for the introduction of the African slave trade to Nueva Granada, which, according to the historians who have investigated the subject, dates back to the second half of the seventeenth century and reached extraordinary proportions during the eighteenth century in the Caribbean port of Cartagena de Indias. This traffic led to the appearance of black ethnic groups in the coastal zones of the Caribbean and the Pacific, and in the Cauca and Cauca Valley regions.

The basic idea behind the new territories was that the king granted dominion over and the right to exploitation of them but not property rights as such (at least in the sense that we understand this today): the latter remained in the hands of the Crown until it decided to award them. In practice, this only happened in the beginning and, as time passed, property rights in line with the traditional models of Spain appeared in colonial legislation. The original form for the granting of rural lands, both in Nueva Granada and Mexico, was that of the royal favours, which were a kind of authorization for the use and dominion of lands that were granted pending the drafting of a definitive legislation for such matters. The favours ("mercedes") were of a transitional character. The intention behind it, which was never fulfilled, was a good one, namely, to limit the rapid formation of great *latifundios* and the concentration of agricultural wealth in the hands of a single landowner. One evidence of this, found in some Neo-Granadian favours, was the obligatory prohibition of their resale or transfer before a 4 – or 6 – year term had elapsed and an absolute ban on their sale to clerics or friars. This, as may be expected, soon became a dead letter. By means of the characteristically Spanish practice of employing front men (or what colonial documents call "prestanombres", that is, name-lenders) – a trick that has been revived in Colombia in recent decades and was used even more widely in colonial Mexico – many neo-Granadian landlords came to own lands that were ten to twenty times as large as the limits permitted under colonial legislation. In Nueva Granada, as in Mexico, the appearance of ever larger *latifundios* was unstoppable from the beginning of the colonial period. What had occurred a long time before and on many occasions in Spanish lands was repeated in the New World, this time on a gigantic scale.

There was no lack of good intentions. The initial ordering of rural space proves it. Confirming what was already the usual practice in Nueva Granada, the "Ordinances for the Discovery, New Settlement and Pacification of the Indies" (1573) state:[13.]

"129. *Mark out for the town enough common land so that, although the population may greatly grow, there shall always remain an ample space where people may go for recreation and cattle may graze without causing damage.*

130. *Bordering on the common lands there shall be marked out pastures for working oxen and for horses and for meat cattle and for the usual number of cattle which the inhabitants, by ordinance, are permitted to have and a good additional amount to be received as common lands of the council and the rest shall be marked out in workable lands that shall be divided into lots according to the quantity that offers itself and in such a way that there be as many lots as building sites to be found in town and if there are irrigable lands make lots of them and distribute them ...and the rest shall remain for Ourselves so that We may grant lands to those who may come to settle at a later date.*

131. *In the workable lands distributed the inhabitants shall afterwards and immediately sow all of the seeds which they can bring ...and in the pastures especially all of the cattle which they bring and may unite so that they may then begin to breed and multiply."*

To briefly synthesize, the Laws of the Indies corroborate the system of ordinances for the countryside as it developed from the beginning of the sixteenth century in Nueva Granada. Starting with the urban perimeter of the town or hamlet and around it, the *ejidos* and *propios,* that is, lands of communal property. Beyond that we find the *suertes* (lots) or farms for agricultural use that were granted in the first sharing-out of lands and then the pastures, of a larger size, for cattle. The rest of the territory, in the more distant parts surrounding the settlement, would be reserved for land grants through royal favour, thus creating properties in which haciendas, in the strict sense of the term, might be formed. Note the insistence, in the Laws of the Indies, on the landowners' taking an immediate and maximum advantage of their properties. This was the limit of the government's good intentions in Nueva Granada, because this royal ordinance was not always complied with. The Colombian historian Carlos Martínez[14.] quotes a piteous petition addressed to King Carlos (or Charles IV) by an eminent inhabitant of Santa Fe (Bogotá), José María Lozano y Peralta, towards the end of the colonial period: *"...so that the rest of those fertile, beautiful and rich lands are villages of Indians and parishes, that is, a church and a priest's house in the middle of lonely and uncultivated countryside...".* Lozano y Peralta must have seen a lot of abandoned or wild land before daring to make such a bold assertion to the king. A short time later the young Colombian nation would inherit the problem of this great and ever-growing distance between socio-political intentions and everyday reality.

The official attitude of the Spanish crown towards the maintenance of the medieval institution of primogeniture, as a factor that might ensure the formation and permanence of *latifundios* in the overseas provinces, was ambiguous. Primogeniture consisted of an express prohibition on subdividing a rural property, awarding its inheritance to the eldest son (or daughter) with the condition that he (or she) must not share out the received properties. Provision number 97 of the

"Ordinances for the Discovery, New Settlement and Pacification of the Indies", issued by Felipe II in 1573, states:[15.]

"To the person who has established a settlement and founded such a town in accordance with Our requirements we award license and authority to create primogenitures or primogenitures for what has been built and in the part of the jurisdiction that has been conceded and in which there are crops and buildings..."

In short, the king made optional, in the overseas provinces, what had been obligatory in Spain but which was already falling into the limbo of customs that had been gradually abandoned. Thus, there is only a partial truth to the hypothesis of the Colombian historian Germán Arciniegas, who writes that one of the basic motivations of the Spanish colonizers, especially those who had the bad luck to have elder brothers, was to escape from the adjudication of family properties

through primogeniture. In the Neo-Granadian countryside primogeniture was only partly applied, during the earliest colonial epoch and in an accidental way, but the tendency to break up *latifundios* and subdivide haciendas located in privileged situations soon relegated that tradition to the limbo of obsolescence or reserved it for occasions when the ruling consideration was convenience or economic interest. A much more important factor in the creation and preservation of *latifundios* were the abundant marriages of convenience among landlords' families, which allowed them to accumulate large properties which they never would have been able to purchase.

What normally happened, as the Mexican historian Ricardo Rendón Garciny and Gisela von Wobeser[16.] point out in *Haciendas de México*, was that the stable haciendas arose as a consolidation of royal favours and, to a lesser degree, of the *encomiendas*. The *encomienda* basically had to do with manpower for agricultural work, while the *merced* (or favour) was a grant of land. The haciendas of Nueva Granada tended to result more from the breaking up of *favours* for land (which were not subject to the legal limitations of primogeniture) than the consolidation of *encomiendas*. By means of the *encomienda* the Spanish Crown entrusted whomever ruled over a particular territory with the management and control of the indigenous workforce, which was obliged to pay tributes in work or kind: the holder of the *encomienda* was also expected to play an evangelical role and impose the Christian religion on his charges. This was simply the formalization of the vassalage to which the natives were subjected. In a certain way, the *encomienda* was a more or less legal method of inserting some kind of official regulation into the relations, rarely peaceful or pleasant, between the indigenous population and the landlords of *latifundios* that arose in the early part of the colonial period.

The formation of the *encomiendas* and, later, of the haciendas, produced a direct clash with the colonial government's intention of organizing and protecting the communities in which the indigenous peoples were artificially grouped. In theory, the natives who lived on such reservations were protected against excessive exploitation by the owners of haciendas and *encomiendas,* but in practice they were also more easily controlled than those who were scattered around outlying districts and it was easier to force them to pay draconian quotas of agricultural or mining labour. More often than not, the reservations were only a residue of lands which had been left to the indigenous communities after the colonizers had deprived them of the areas best suited to agriculture and housing. This confrontation, in the countryside of Nueva Granada, was not a a matter of adjusting a Spanish system of landholding and agriculture to an indigenous one. The uprooting of the Indians and their grouping, as peasants, in new communities headed by a *corregidor* (chief magistrate) named by the colonial administration was a strictly Spanish idea.

Margarita González[17.] (op. cit.) points out that when the President of the Royal Audience, Andrés Díaz Venero de Leyva, issued, in the middle of the 16th century, "*...the first provisions for the establishment of reservations ...The reaction of the holders of "encomiendas" was so negative that the Royal Audience had to think about postponing the creation of the reservations...*". According to historians who have studied this subject, the reservations "of Indians" rarely or almost never won a lawsuit or argument about lands, tributes or labour quotas brought against the haciendas "of Spaniards".

The practical and legal disorder that characterized the early history of the Neo-Granadian colony made necessary the successive practice of *composiciones* (settlements), by means of which the landlords standardized (and legalized) the system of land deeds and taxes agreed on with the colonial administration. In *Popayán, una sociedad esclavista,* G. Colmenares[18.] writes that: "*...From a judicial point of view the 'composición' was nothing more than a way of acknowledging the sovereignty of the monarch over the conquered lands. Although, in principle, a de facto situation was recognized ...the Crown refused to acknowledge the legal validity of such 'mercedes'* (granted by town councils or Governors) *since it had never relinquished the faculty of granting them. This gave it a pretext to demand a payment... for the validation of deeds arising from previous grants*

Opposite page, House in the sierra of Santa Lucía, province of Cádiz, Andalusia. An example of how the Andalusian countryside was settled with a minimal architecture.

or based on mere possession". These settlements began in Nueva Granada, according to Colmenares, at the end of the sixteenth century and the last one occurred starting in 1637 in Cauca and the Cauca Valley, with Popayán as the centre.

Paradoxically what limited and put obstacles in the way of rural work and development in Nueva Granada was the very nature of the colonial economic system. Except for regions where agricultural production left certain surpluses or capital reserves (such as those resulting from the production of sugar cane and its derivatives in the Cauca Valley) there did not exist any mechanism of credit other than the ferocious usury practiced, more or less in secret, by private individuals (usually merchants) or the *a censo* (mortgage) loans provided by religious communities, whose incomes allowed them to carry out this activity. It is worth noting that some of the religious orders in Nueva Granada were miserably poor and depended on public charity to survive, while others were so rich in income and capital that their situation was resentfully remarked upon by their neighbours and contemporary chroniclers. The *censos,* in theory, were low (for the period) interest loans but, in practice, due to the permanent economic difficulties or bankruptcies of numerous landlords, their rates were much higher than the legally permitted 4 or 5% And even at this interest rate those who recurred to the *censos* were often unable to pay the accumulated interest. Thus it was that the religious orders, especially the Jesuits, gradually took possession of estates and farms whose owners were unable to pay the taxes on their lands and houses owed to the colonial authorities. The uncertainties and fluctuations of an agricultural production that had a mediocre technical level and was poorly administered, together with the proverbial bad luck of farmers, helped to keep the Neo-Granadian countryside from becoming an emporium of wealth. According to G. Colmenares[19.] (op. cit.), by the eighteenth century religious orders, as a group, had probably become the single most powerful and productive block of landlords in Nueva Granada. Perhaps there is more truth than one may imagine to a well-known piece of gossip of the period, which was recorded by a number of chroniclers – namely, that before they were expelled from the province of Nueva Granada, the Jesuits owned so many cattle ranches (especially in the Eastern Plains region) that the inhabitants of Santa Fe (Bogotá) could only eat beef when the Jesuits allowed them to. While there is no doubt that there came to exist a small number of hacienda owners whose incomes provided them with a comfortable economic stability and certain small luxuries, even the most remarkable rural fortunes in Nueva Granada were unimportant beside the astonishing accumulation of wealth that took place among Spanish, Portuguese or Creole landlords in Mexico and Brazil. While a small number of mining estates did exist in Nueva Granada, they did not show (and never could have shown) the fabulous profitability of those of Peru or Mexico. The enormous agricultural development of Nueva España (Mexico) in the colonial period, backed up, as it was, by the large-scale mining of precious metals, seems to belong to a continent–or world–that is completely different from Nueva Granada. It was, in a way, the *other* Spanish empire.

In view of all this, the slow and deliberate growth of hacienda houses in Nueva Granada, as well as their architectural modesty and humbleness, are understandable. During the process of taking possession of (and exploiting) the countryside, the building of a house which reflects the owners' social standing and fulfills the functional requirements of rural life only comes about after other and more practical needs have been attended to, such as the organization of the workforce, the providing of hydraulic resources, the obtaining of capital reserves and the creation of a system for dealing with administrative and tax matters. If we add to this the fact that the landowner owns a "main" house in the nearest city or town, we can understand why the building of a hacienda house would have had a relatively low priority, being considered almost as a luxury.

According to the Mexican historian Ricardo Rendón Garcini[20.] (op. cit.) "*...the haciendas did not appear in a given moment or on a specific date. Nor was it a system of production that had been theoretically preconceived* *and then brought into practical being through a legal decree".* This is clearly true for Nueva Granada. The turning of lands received through a royal favour into a consolidated and stable hacienda sometimes lasted the better part of a century, that is, four generations, starting at the beginning of the conquest and finishing with the colonial *composiciones.* The appearance of a great hacienda house in Cauca, Boyacá or the savannah of Bogotá, built at the same technological and qualitative level as its urban equivalents, would take even longer, that is, not until the end of the seventeenth century or the beginning of the eighteenth. Germán Colmenares, in *"Popayán, una sociedad esclavista"*[21], significantly observes how, in the inventories of the haciendas of Cauca in the seventeenth century (conserved in the Central Archive of Cauca), the slaves, farm equipment and work tools had a value that was much higher than those of the rural houses that lodged them, a situation that did not essentially change even in the eighteenth century.

The Neo-Granadian hacienda house is a slow and gradual consequence of the historical formation of the hacienda itself and, the same as the fruits of the earth, requires that the species which engenders it ripens little by little. It has been mentioned above how, in the places chosen for building in the countryside, a process of successive improvements had to be gone through, starting with rough shelters or huts improvised with whatever materials were at hand, and then, if all went well, the first outline of what, with the passing of time and by knocking down here, adding on there and rebuilding in general, in accordance with the fortunes of the family or the socio-economic conditions of the region, would become a finished hacienda house. This, in short, would be the generic history of the Neo-Granadian hacienda houses. Of course, there were also houses which were erected, all at once and almost completely, in places where no building had been done before, but this was exceptional. Such cases are more characteristic of the republican period (the nineteenth century) than the colonial one.

The author of the present essay, in previous studies of the Neo-Granadian hacienda houses, made some observations about the anthropological and social aspects of their construction and the management of the surrounding countryside, which may be summarized in the following way: *"Rural life in Nueva Granada was in a minor key."* The Spanish thinker José Ortega y Gasset says, in his Theory of Andalusia: 'The Andalusian lives in a fat, exceptionally fertile land which gives fruits with a minimum effort ...Like a plant he receives nourishment from the earth and receives the rest from the warm air and the beneficent light...

"Culture does not consist of anything other than finding an equation to solve the problem of life. But the problem of life may be set forth in two different ways. If, by life, we understand an existence of maximum intensity the equation will force us to confront a maximum effort. But if we previously reduce the vital problem, we may aspire to a minimum life. Therefore, with a minimum effort we will obtain an equation as perfect as that of the most heroic nation: this is the case of the Andalusian... Instead of increasing what he has, he reduces what he needs; instead of exerting himself to live, he lives so as not to exert himself. He makes the avoidance of effort the principle of his existence". The logical transplanting of the Andalusian equation to Nueva Granada endorses the theory of Ortega y Gasset. The choosing of certain regions of the province for agrarian activities, finding the most benign climates and avoiding those which would have required titanic efforts to establish a certain productivity, are other evidences that corroborate the Andalusian attitude described by Ortega y Gasset. Thus, the Neo-Granadian hacienda house could never be less than a profound reflection of this reason for and way of being. There was no reason to exert onself (for what?) in erecting enormous country palaces or vast rural citadels; there was no need to work desperately, from dawn to dusk, in lands which produced three or four harvests in the same time that was required to produce one harvest on Spanish soil. It was vital to reach an agreement with the new environment, forgetting all that had to do with the annual seasons and trying to adapt oneself to the difficulties of the tropics or the high moors, instead of attacking them head on. If the rural life of Nueva Granada was muted and low-

key, so would be the architecture that it lodged: humble but not insignificant. Formally modest, but with a profound metaphysical dimension. Reticent, but the unmistakable lord of the surrounding landscape.

The Andalusian attitude towards the countryside was not the only one that would make itself felt in New Granada. The colonization of rural Nueva Granada was by no means the work of a single provincial group. It is not even certain that the Andalusians were a majority in the region. Settlers of low or middling social strata from Castile, Extremadura, Galicia and even the Basque Provinces reached Nueva Granada and it was only later on that a few wealthy merchants or aristocrats in search of a better way of life followed them. Only those from Catalonia were, by royal will, considered to be foreigners in the colonies, their mass immigration to Colombia mainly taking place during the nineteenth century or towards the end of the eighteenth. The tough nature, endurance and stubborn approach to adversity which all of them displayed were necessary for the colonization of the countryside of Boyacá, the mountains of Antioquia and the Eastern Plains. This mixture, of course, does not weaken the validity of the "Andalusian equation". It was possible to meet the challenge of imposing a European agrarian culture on the countryside of Nueva Granada without altering the terms of the *minimum life* and without adopting an attitude of superhuman effort.

The answer given by anthropologists and sociologists to the question of who settled and built in the countryside of Nueva Granada revolves around two basic points. The first consideration has to do with the groups of Spanish settlers of diverse regional origins who reached Nueva Granada and took (or usurped) lands whose property would have to be consolidated in accordance with European hereditary traditions. And the second, that, while the great and middling Neo-Granadian landlords were originally Spaniards, they were succeeded by creoles, that is, "white men" or descendants of Spaniards who had been born in Nueva Granada and who, for reasons of social status, had a relatively easy access to agrarian property. We shouldn't lose sight of the extraordinary socio-economic and even psychic transformation that took place among the Spanish peasants or *petit bourgeois* who were lured to Nueva Granada because of the abundance of land that had been seized or shared out there.

The peasant turned into a *hidalgo* (nobleman), as confirmed by the Laws of the Indies. The bourgeois shopkeeper or mason became a respectable gentleman. Along with the land came pride and vanity, at times excessively so. One understands the protest made by Felipe IV of Spain when he angrily asked who were those Neo-Granadians that, from the other side of the ocean, dared to address themselves to him with the title of "Don" before their Christian names, without having the courtesy to ask for the authorization of their Monarch. A great Neo-Granadian landlord might possess twenty times more land than his Andalusian equivalent, but he would never get the latter to acknowledge that they belonged to the same social class. In its overseas provinces, Spain had created what it considered to be a permanent species of upstarts.

Nueva Granada could never been been the setting for the singular phenomenon of those indigenous "nobles" who, as chiefs of tribal groups, came to possess (or continued to hold what had always been their own) vast tracts of land and turned into powerful hacienda owners in the centre and south of that country, rivalling Spanish landlords. By contrast, there took place, in Nueva Granada, an extraordinary crossbreeding, which spread across racial frontiers and diluted the presence of indigenous groups in the countryside. A great role in the process of racial mixture is justifiably attributed to Spanish and, later, Colombian landowners. Even today, in some regions of Colombia, people still repeat an old joke, from the Colonial period, about this phenomenon. It has to do with a hacienda owner who rides over his vast properties on a purebred Andalusian horse and meets up with a boy whose face is neither dark nor white and whose features show neither the angularity of the Indian nor the "purity" of the Spaniard. In response to the eternal question of who his parents are, the boy quickly and smilingly replies: "Well,

Opposite page, *Los Aposentos*, Chocontá, Cundinamarca.

I am the son of daddy Manuel (or Pedro or Sebastian or whatever the name of the rider is...). "And then comes the sharp reprimand: "Don't play the fool! That I already know. What I am asking you is, which Indian is your mamma?"

In Cauca and Boyacá, during the Spanish colony and then at the beginning of the republic of Colombia, illegitimate descents, whether recognized or not by Indian or national law, rapidly turned into extraordinary and abundant genetic pyramids. In the countryside, as is well known, such customs as the celebrated and infamous colonial *derecho de pernada* (the equivalent of the feudal *droit de seigneur)* had no other limit than faint cultural conventions or the simple and crude physical ability to get away with it. The racial mixture was necessarily going to include the descendants of the African slaves brought to Nueva Granada starting in the seventeenth century, which would produce more than one eighteenth – or nineteenth – century hacienda owner and family head who boasted of a Hispanified Arab surname, spoke with an Andalusian accent, said that he was a descendant of a *conquistador* but displayed unmistakably negroid features.

There would also come into being a peasantry abounding in landless serfs but, at the same time, mestizos gradually and with difficulty gained access to land ownership, depending on the socio-economic situation they had reached. In the countryside the advent, in the colonial period, of a sort of "middle class" that owned land but was less wealthy and powerful than the great hacienda landlords and, of course, without the racial purity which, for example, the oldest families of Popayán, Santa Fe or Tunja might claim, occurred, above all, during the last century of the colonial period. Among other things, when almost all of the rural landholdings of the religious orders disappeared in the nineteenth century, somebody had to take their place and it was not always the richest hacienda owners who benefited from the laws prohibiting religious orders to own rural properties. The coexistence of rigidly stratified social classes was more viable in the country than in the cities, where the "aristocracy", that is, "respectable" people accepted the presence of "the rest" with reluctance or not at all. The Colombian (or republican) period was thus brought into being, when the dominant presence in the countryside would be a dual one: the owners of great estates, on the one hand, and on the other, middle class hacienda owners, merchants, small industrialists and those who had some capital – a group that was much more numerous than the former and, as a whole, owned an enormous amount of land, even though the individual properties were smaller than those of the great landlords.

The question which obviously arises here is, what or how much did the Neo-Granadian hacienda and *latifundio* landlords actually own? It is impossible to compare the possessions of the Count of Las Torres in the northern Mexico, which were thought to be as large as Scotland and part of Ireland, or those of some Brazilian landowners, whose properties might have reached the order of a fifth of the whole of Nueva Granada, with what might be termed a *latifundio* in Neo-Granadian terms. Even so, there certainly existed haciendas in the regions of Cauca, Huila and the Eastern Plains that were so large that their colonial owners never had more than a vague and poetic idea of their size and boundaries. Among the great estates that are documented we find, for example, La Dehesa de Bogotá, in the savannah to the south-west of Santa Fe, which was one of the few which really deserved being called a *latifundio*. Although the celebrated Dehesa was the object of numerous legal transactions which altered its original size and shape, we can trace its origin back to an *encomienda* that passed from the hands of Luis Alonso de Lugo to those of the royal ensign-bearer, Antón de Olalla. Lugo is known for having brought the first bulls and cows (35 of each) from Spain to the high plains of Santa Fe, that is, he was a pioneer among the Spanish owners of haciendas devoted to cattle-raising. Although it is difficult to establish the precise boundaries of La Dehesa, it probably covered the present-day municipalities of Funza, Serrezuela, Mosquera, Soacha and parts of Bosa, Fontibón and Facatativá, its approximate area being, at some point in time, some 45,000 hectares, which is somewhat larger than *the current urban area of Bogotá,* without including the adjacent towns that are annexed to it.[22] By the middle of the eighteenth century a complex process of subdivision had reduced the Dehesa to some 12,000 (or 14,500) hectares, including the haciendas of *Fute, Aguasuque, Canoas* and *El Novillero.* It was in this state that the Marquis of San Jorge, don Jorge Miguel Lozano de Peralta y Varáez Maldonado de Mendoza y Olalla received the Dehesa as an entailed estate, a futile land deed insofar as the dismembering of it was already well advaanced and would continue into the nineteenth century.

It is significant that, in the second half of the eighteenth century, the biggest landowners in the south-western zone of the savannah of Bogotá, apart from the Marquis of San Jorge, were the nuns of Santa Inés, although the nuns of Santa Clara, Augustinian Fathers, Dominicans and Jesuits also owned the best lands of the region. The nuns of La Presentación (Santa Inés) were the owners of the haciendas of *El Salitre, Cortés* and *Las Monjas* on the right bank of the river Bojacá, as well as *El Corzo, La Jabonera* and *Serrezuela,* some of which, although much reduced, still bear their original names. It is possible that the properties of the nuns of La Presentación in this part of the savannah of Santa Fe alone amounted to something more than 4,000 hectares.

According to the data of Germán Colmenares (Las *Haciendas de los Jesuitas en el Nuevo Reino de Granada),* the Jesuits came to own a total of 72 haciendas throughout Nueva Granada, including those of *Techo, La Chamicera, Tibabuyes, Fute, El Chucho, El Vínculo* and *La Calera* in the savannah of Santa Fe and its surroundings.

The areas of the haciendas and farms (*Quintas,* as they were called around Santa Fe) in Nueva Granada as a whole varied a great deal and, in some regions, it is difficult to say whether they had a standard size. In the Cauca Valley, according to G. Colmenares[23], there came to exist *latifundios* as large as 34,000 hectares, such as that which was owned by Antón Núñez de Rojas in what is today the Municipality of Cerrito, although in the same region the lands of the hacienda *El Alisal* reached an area of no fewer than 6,000 hectares in the eighteenth century and there are many recorded transactions, in the surrounding zone, that refer to the sale of "pastures" or farms of between 30 and 300 hectares. Colmenares points to a similar situation around Popayán, where, in the middle of the eighteenth century, the process of dismembering and rebuilding haciendas was a complex one, due to the very widespread practice, by hacienda owners, of monopolizing lands through property deals and marriages of convenience.[24] One singular evidence of this, according to Colmenares, is the fact that the number of rural landowners in the jurisdiction of the city only increased by 15 between 1637 and 1713, a period of 76 years! In short, the area around Popayán was in the hands of a very small number of landlords, among whom two women were highly prominent, Josefa Hurtado del Aguila and Dionisia Pérez Manrique, the Marchioness of San Miguel de La Vega. The Marchioness was a typical personality of the epoch. The daughter of landlords, she first married José de Velasco, the holder of *encomiendas* who, when he "passed on to a better life", left her enormous properties; she then married the Marquis de la Vega, the owner of gold mines in the Chocó, which made him a rich man in Popayán circles. When the Marquis conveniently died, his widow was able to count, among her possessions, an accumulation of land that was equivalent to twenty times the urban area of Popayán.

The end of the colony also witnessed the speeding up of a process of splitting up of the large-scale *latifundios* and haciendas – it was the last act of the "golden age" of colonial houses. New social groups that possessed a growing economic power would come to completely change the regime of rural land ownership, breaking the monopoly of a few landlords as the country entered into the republican era.

Left and above, The house of *Britalia,* Duitama, Boyacá. The hacienda it belongs to was established at the beginning of the eighteenth century, although the house itself may be more recent. It conserves its original spatial organization and volumetry, but the beautiful and gentle landscape of its original setting is suffering a process of deterioration and destruction due to the growth of the suburbs of Duitama. Taking advantage of the sloping hillock, *Britalia* includes a short ground floor upon which the gallery of the main face is placed, a feature seen in other houses pictured in this book. This allows it to be used as a storeroom and achieve a more spacious volumetry. The depressed arches, like the unusual enclosed balconies that are characteristic of the urban houses of colonial Santa Fe (Bogotá), are symmetrically placed upon the façade and result from nineteenth or early twentieth century reforms. The inner courtyard or patio retains the original constructive system of wooden pillars and lintels, but the stone fountain is a modern decorative detail.

Left and opposite page, *Tequendama,* Chusacá, Cundinamarca. It dates from the eighteenth century, although the house has undergone several extensive transformations, including the typical republican enclosed upper galleries towards the inner courtyard and façades. The hacienda, like many others on the savannah of Bogotá, originated from a royal land grant of the eighteenth century.

"…the worst thing that can happen to a hacienda house… is the destruction or defiguration of its setting or immediate surroundings. This affects the most important characteristic of the hacienda house, which is the relation between the house and the natural space which surrounds it. The cutting down of ancient trees and the demolition of surrounding walls are the main examples of vandalism…"
G. Téllez

Opposite page and right, House of *Suescún,* Tibasosa, outskirts of Sogamoso, Boyacá.

Opposite page, above, The complex of enclosure walls of *Suescún,* one of the most beautiful in the high plains region of Cundinamarca and Boyacá, in 1962. This admirable transcription of the Andalusian concept of successive spaces has nearly disappeared.

Opposite page below, The front of the principal section of the house, with its compact floor plan.

Right, The most important space in the house: the gallery facing the garden, from which it was possible to glimpse the spaces defined by the enclosure walls and the distant valley.

Above, right and opposite page, the house of *Suescún,* Tibabosa, outskirts of Sogamoso, Boyacá. Located in what, only a few decades ago, was an idyllic spot in the valley of Sogamoso. The hacienda itself, which resulted from the breaking up, in the first half of the seventeenth century, of a vast estate which covered the whole of the surrounding area, was gradually reduced in size to the few hectares which now compose its setting. Tourism has saved it from disappearing or being gradually disfigured, but it is still threatened by factors of economic and industrial "development", such as atmospheric contamination, the near presence of urban housing projects, a local cemetery, etc. For this reason, it is a miracle that *Suescún* still conserves part of its agreeable environment and the recondite architecture of the house. *Suescún* does not offer a spectacular architecture or special effects. Rather, the house is content to be in harmony with its setting in a simple and intimate way. It modestly stands on a slight rise and looks out on the valley beyond its walls, which, for the most part, are now only preserved in our memory.

Right, Given the circumstances of the original landscaping, the gallery of the house turns out to be a masterpiece of rural architecture.

He who is imprisoned in this tower would stay imprisoned.
(Four windows to the wind).
– Who screams towards the north, my lady?
– The restless river.
(Three windows to the wind remain).
– Who howls towards the south, my friend?
– The air, which blows tirelessly.
(Two windows to the wind now).
– Who sighs towards the east, woman?
– Yourself, who comes in death.
(And now one window to the wind)
– Who weeps to the west, my friend?
– I, now dead, who go to your burial.
Nothing would induce me to stay imprisoned in this tower!
Rafael Alberti.
The Tower of Iznájar.
(1926).

Right, What is unusual in the house of *Suescún* would be commonplace in many parts of Andalusia and La Mancha: a tower resembling both a watchpost and a belfry, erected sometime during the seventeenth century by a hacienda owner who wished for a more distant view of the landscape ...or to be nearer to the heavens. This ingenuous architectural feature lends a Spanish touch that distinguishes *Suescún* from its Neo-Granadian fellows.

Left and above, The house of *Suescún,* Tibabosa, outskirts of Sogamoso, Boyacá. Interior of *Suescún.* The house has been a tourist hotel for the past forty-five years, which explains the peculiarities of its current atmosphere. The round rafters used in the roof frames with auxiliary struts or braces, a typical feature of the region, have been conserved during recent alterations, but the chimney, which evokes a traditional baking oven characteristic of the kitchen area, is a modern element that has been installed in the reception rooms.

Left, opposite page, above and below, *La Concepción de Amaime,* El Cerrito, Cauca Valley.
This is, in architectural terms, the most singular and outstanding hacienda house of the region. It reached its "final" form in the second half of the eighteenth century. Its volumetry within a flat and topographically undistinguished place stands out, which is why it relies on its own architecture to define the spaces around it.
In addition to the landlord's house, with a compact ground plan, it has a free-standing chapel, a shed that served as the peons' quarters and storerooms and some remains of a large aqueduct which supplied water to, among other things, a now disappeared traditional open-air bath. A sugar-cane mill is to been at a certain distance from the house. The location of the chapel and the raised volume of the western end of the house, which is provided with an upper balcony and a lower gallery with sculpted columns, form a splendid architectural whole.

Above and right and opposite page, *La Concepción de Amaime,* El Cerrito, Cauca Valley.
The exquisite volumetry of *La Concepción de Amaime* is an architecture of masons and carpenters, but it is also a masterpiece in terms of its skilful use and combination of regional building resources, an example, that is, of artisanal design in the best sense of the word. Only a few Neo-Granadian hacienda houses of the eighteenth century invest strictly technological considerations with the poetical quality that is needed to raise them to the level of great architecture.
In contrast with the general run of regional architecture, it is the house of *La Concepción de Amaime* which imposes its presence on the place and grants it a special atmosphere. The utilitarian need to achieve a visual command of the surrounding territory, that is, the world which surrounds the house, was resolved, in *La Concepción de Amaime,* by means of a splendid running balcony which, as B.Barney and F.Ramirez indicate, in their study of the haciendas of the Cauca Valley, evokes the colonial architecture of Cartagena.

Right, The free-standing chapel is a modest volume of a single nave, but to this was added, in the eighteenth century, a mildly baroque wall belfry and altarpiece and a façade decorated with imbedded cornices and columns, as well as a certain degree of decorative embellishment around the arch over the main door, luxuries that are unusual in the oratories or chapels of hacienda houses.

Left, above and below, *Gotua,* Iza, Boyacá. It may have been built in the second half of the eighteenth century or the beginning of the nineteenth, when the large colonial estates which covered the valleys of Sogamoso, Firavitoba, Iza and Pesca began to be subdivided. It is a typical low Boyacá house, organized around two, three or (at times) four sides of a central patio. We may classify it as a minimalist rural architecture, that is, not much more could have been done with the modest resources at hand. Like other of its equivalents in the region, it was subjected to alterations in the republican period, when very low ceilings were installed and door and window openings were equipped with "kneeling" iron grilles characteristic of nineteenth century urban houses. The atmospheric emphasis of *Gotua* is almost entirely upon the interior courtyard, whose proportions and character strike an evocative and pleasing note. Each patio of this kind in Boyacá has its own individual features, even though all follow the same overall principles of spatial order.

Left and opposite page, *Gotua,* Iza, Boyacá.
The house of *Gotua* is an example of the poetry of the functional. A corridor or gallery is no more than a space that one goes through. A portal no more than a point of access. They may be banal and indifferent or evocative and unforgettable. The modesty of *Gotua's* patio and entrance has not been altered by pretentious brick arches and stone columns which condemn to oblivion the simple columns and lintels made of local woods and decorated with gay colours.

Previous pages, opposite page and right, *La Merced,* El Cerrito, Cauca Valley. Just as in other houses in the Cauca Valley, the environmental character of the site is determined more by the splendid vegetation than any topographical feature. *La Merced* includes an original upper storey built with the same idea of providing a a commanding view of the surroundings that was mentioned when we spoke of *La Concepción de Amaime.* In contrast with the latter, *La Merced* conserves a good part of the walls or adobe bricks that join the landlord's house to the peons' quarters and other outbuildings, creating a remarkably interesting volumetric whole. The conservation of the house as an architectural grouping gives it an important place in the country's architectural heritage.

Opposite page and right, the main galleries of *La Merced* include, like those of *La Concepción de Amaime,* the unusual luxury of columns made from extremely hard wood from the Chocó region, carved with a certain artisanal fantasy. Such an architectural effect is exceptional in Neo-Granadian rural architecture.

Above and right, *Casa de Teja,* Guasca, Cundinamarca. Built towards the end of the eighteenth century or the beginning of the nineteenth. As well as its charming Andalusian nickname, which is common in the south of Andalusia (where it is used to distinguish rural buildings with tiled roofs from those with flat ones in terraces), the house conserves a rural version of a large roof frame done in the "par y nudillo" techniqe and in which even the transversal tie-beams and the "cuadrales" (diagonal pieces in the corners) are made of round logs, with a "chusque" (light Andean bamboo) support for the tiles. Like *Gotua* in Boyacá, *Casa de Teja* is an example of architecture that has been reduced to the essential, in a minimalist style which would otherwise lack atmospheric charm.

Opposite page, and right, above and below, *Chaleche,* outskirts of Sesquilé, Cundinamarca. Three houses exist which bear or have borne the same name: the late-colonial hacienda house shown here and two interesting republican farm houses built between the end of the nineteenth and the beginning of the twentieth centuries, products of the subdivision of lands of the colonial hacienda, which, in turn, resulted from the breaking up of the vast land-grant estate of Sesquilé at the beginning of the eighteenth century. The house from the colonial epoch seems to have grown, by stages, until well into the nineteenth century, without undergoing brusque changes of volumetry, such as the upper floors added to the Cauca Valley hacienda houses. This explains the unassuming relationship between the landscape and the present form of the house.

Space, design and technology

"A region which history has not marked with the passage of its deeds, which literature has not embellished by including it in its pages... which painting has not brought to its canvasses will be a series of haciendas in which the cattle become more or less fat and the landlords proportionally wealthy; but it will never have the sharply-defined features that impress the minds of men, nor lead their spirit to the inner contemplation of the profound richness of the past and the mystery of the future."

TOMAS RUEDA VARGAS.

La Sabana. 1940.

Perhaps the most complex but strongest link between the countryside and the hacienda house is what is known as the sense of place. In *Casa Colonial,* the author of this essay wrote:

"The sense of place is a matter of ...mysteriously extracting certain meanings that are implicit in the natural space and creating, with these, an harmonious artificial shaping of the environmental elements ...It is not enough to impose a certain architecture upon a place; the house should be as efficient as a farming tool and, at the same time, look as natural as the landscape which surrounds it. An intangible but not imperceptible dialogue has to be established between the place and the presence of its architecture."

In the first chapter of this book we mentioned how the memory of and analogy between landscapes was part of this sense of place for the Spanish settler and builder, but this is only one component in the perceptive and creative process through which the choosing of a site and of a system of artificial forms provides a physical framework to the ownership of a particular part of the countryside. The sense of place which leads the Spanish or creole peasant or bourgeois to behave in a certain way before the landscape which he chooses for himself has two main aspects: the first is a pragmatic or functional one, since without a supply of water, an access road and a reasonable availability of building materials, the problem of the hacienda house would never arise. If it was not possible to site the house in the immediate vecinity of a well or a stream, or it was more prudent to place it a distance from the bank of a river where floods were a danger, there was always the possibility of improvising some form of aqueduct or even using a home-built version of the ancient Roman hydraulic ram or Arab waterwheel to carry water uphill. Where sources of water were not evident, one could recur to those, among the settlers of Nueva Granada, who possessed the gift, science or witchcraft of water-divining. With its infinite variety of landscape and climate the countryside of Nueva Granada was divided, like Castile or Andalusia, into regions of dry land, where the owners of *encomiendas* and haciendas had to depend on tremendous tropical rainstorms and regions of irrigable land or those with more or less permanent sources of water.

The second aspect, which necessarily had to do with the first, would be the poetical one. Among the numerous definitions of this word found in the Dictionary of the Spanish Language we find: *"...a group of qualities which must characterize the background of this genre of the product of human understanding, independently of the external form...".*

The heart of the matter lies in the fact that, from a complex of pragmatic considerations, there arise, in the case of the Neo-Granadian hacienda house, a discreet and profound aesthetic dimension and a certain environmental grace, tokens of an ancient and noble origin which are inherent to it and not academically superimposed, as occurs in much of "cultured" or monumental architecture. It is clear that a low-keyed architecture, devoid of spectacular formal gestures is, by definition, more in harmony with any landscape and place, since it does not pose intolerable contrasts. It is precisely with this second aspect that emotional memory plays a decisive role. Each and every hacienda house, following the definition of the Spanish historian Maricruz Aguilar[25], is so intimately linked with the natural landscape where it is located that it is difficult or impossible to imagine the landscape without the house or to think of the house in another place. The essential part of the definition given by the Dictionary is the way the "external form" becomes independent, which has largely to do with practical considerations. The harmony between or integration of the natural space and constructed forms in the genre of hacienda houses is not regulated or established by abstract theoretical concepts or "golden rules".

Opposite page, *Aposentos,* Simijaca, Cundinamarca.

By choosing a particular spot in the countryside and saying, "we will build the house here. Not over there or farther away but here", the landowner summons the poetical to guide the future destiny of the place, calling upon the four cardinal points, the winds and rains of the year's seasons, the sun and the phases of the moon and, without knowing it or through a blind intuition, the spirits of the earth and the air, like a sorcerer or shaman. The beautiful location of the house in the Cauca Valley which was originally called "de la Sierra" ("of the mountain") confirms this. In the novel which he set there (*María*) Jorge Isaacs gave it a name which does not refer to the place where it is situated but to the romantic lyricism of the epoch: *El Paraíso* (Paradise). That this ideal was never reached in real life is another matter, which stems from the inevitable gap between the all-too-human fallibility and the idealism of the Neo-Granadian landowner.

Part of the complex phenomenon of perceiving natural or artificial space has to do with the establishing of physical dimensions which impose an order and, in a certain way, make it tangible. One thus creates what scholars have called the mathematical space. But it is also possible to arrive at a perception of space by means of ergonomic experience (physical gestures and actions) in what anthropologists call the vital space.

The sum of vital experiences occurs within what might be called the existential space. If a hacienda owner in the colonial period has to spend a whole afternoon to travel, at the tired pace of his mule, from his estate house on the outskirts of Facatativá to Santa Fe, covering a distance of some five and half leagues, the mathematical space he travels would be exactly those five and a half leagues; his vital space, that which is occupied by the saddled mule and rider, including his poncho, hat and staff; and the existential space, a sort of imaginary tunnel through the cold air of the savannah of Santa Fe that is spread along the whole length of the proposed ride, within which one would insert the path and tired gait of the mule, including the halts to drink corn (or sugar cane) liquor in the various taverns along the highroad.

Colonial surveying, like colonial construction, was based, not on systematic, but on unsystematic notions of measurement, partly supported by traditional anthropometric tools. The use of human measures and ergonomics that could be multiplied, increased or combined, a tradition derived from Greece and Rome, allowed the Spanish landowners to attempt to impose measurements on the immeasurable Neo-Granadian landscape and then lay out the dimensions of their houses. Thus, the *vara* (yard), in building, and the *paso* (step), in the countryside, were, as the folk saying has it, "the same, only different".

The *vara* was an abstract measure, made up of three lengths of a human footprint or by four palm-lengths (taken between the little finger and thumb of an open hand). The *paso*, measured between heel prints on damp ground, was equal to a *vara*, although some historians think that it was equivalent to two-thirds of a *vara*.

In the natural space, the *paso* (also called *vara de campo*) was the minimum unit. To facilitate measurement a multiple known as a *cuerda* or *cabuya* (cord) was used, whose length arbitrarily varied between 64 and 120 *varas*. In Nueva Granada, the word *estancia* was not used as a generic term for hacienda but meant a unit or measure of area. Although this varied a great deal from one region to another, Germán Colmenares, in cali: *Terratenientes, mineros y comerciantes*,[26.] notes the tendency, in the Cauca Valley, to use the traditional unit of the *estancia,* that is, 6,000 by 6,000 *pasos* (a bit more than 500 metres), for measurements on cattle farms. With the passing of time (towards the end of the seventeenth century) and as this unit became impractical for their size, the typical *estancia* was reduced to 3,000 and 1,500 *pasos*. The *estancia* used for sheep and pig plots began by being 3,000 by 3,000 *pasos* and later showed the same reduction applied to that of cattle farms. The *estancias* used for crops had an area of about 2,000 by 1,6000 *pasos* (some 168 by 134 metres). In the region of Cauca, by contrast, the "official" system of agrarian measurement, established in the Laws of the Indies and based on *caballerías* and *peonías*, applied, although Germán Colmenares points out, in *Popayán: una sociedad esclavista,*[27.] *that "...the units of mea-*

surement were highly anarchic. Not only were a great number of them employed -although some were prevalent in certain areas - but the same term covered different concepts ...It turns out to be almost impossible to establish what a 'caballería' was unless we adhere to the definitions or practices used in a particular place". Colmenares shows how the size of a *caballería* in Nueva Granada might vary between the 427 hectares used in the surroundings of Cartagena to the 250 found in the Cauca Valley. The *peonía* designated, in principle, lands smaller than those of the *caballería* (that is, about a third or quarter of the latter), destined for crops. According to the same author, the most widely spread agrarian measurement in Nueva Granada was the *hanega de sembradura,* which did not have an homogeneous area either, varying, according to the region, between a bit less than two to nearly four present-day hectares. We can appreciate the worrisome anarchy of colonial surveying by mentioning that other measurements derived from the *caballería* and the *hanega* or *fanegada de sembradura,* like the *suerte de tierra* (1/4 or 1/6 of the *caballería*), which means to say that the *suertes* did not have a uniform area either. In the upper Cauca the ancient *legua* (league) came to be used, although interpreted with absolute freedom, for the granting of royal favours of a great size. The *legua quiteña* was the equivalent of about 560 hectares and was different from the *legua tirada,* a straight-line measurement used for rural distances and usually denoting a length of one hundred blocks of a hundred *pasos* (8.4 kilometres).

This delightful anarchy was part of the sensory impact, on the Spanish settlers, of the landscape and geography of Nueva Granada, for how could one measure the prodigious dimensional scale of the New World? There was no place in the whole of Spain where one could say that the boundaries of a single hacienda lay at a distance of three or four days'journey on horseback. Or that the distance to sources of water was "three cigars on muleback", that is, the time it would take for the mount to reach that place at a slow stride, which was the exact equivalent of how long the rider would take to smoke three "butts" of the magical tobacco of the New World, keeping the lit end in his mouth so that the wind wouldn't blow ash into his eyes.

The building of a hacienda house belongs, and honourably so, to the noble class of the inexact sciences. In the absence of some unknown abstract dimensional system, building was done on the basis of anthropometric measures. The tallest person among those present was chosen: he took a step forward and the distance between his heel prints was measured. Then a thin rod of the resulting length was cut and smoothed off and you got what was called a *vara de la tierra* (a land rod), which was different in each hacienda and village. The Spanish crown tried, in vain, to tidy up this happy informality by ordering the standardization of what was called the *Vara del Rey* (King's Rod), which was closer to the English yard than the *vara de la tierra.* But in this case as well the colonial norm – to formally obey but not practically comply with – applied. It is not easy to determine the exact metric dimensions of the *vara* insofar as the measurement depended on local interpretations and convenience, but among the "King's Rod" or the "Castilian Rod" (another traditional version) and the many lengths of the "land rod" there might have been a variation of up to 9 centimetres (between O.83 and O.92 metres), with the most usual average being between 0.846 and 0.867 metres. The *vara,* of course, is conceptually related to the English yard and was divided into halves, thirds (feet or elbows), quarters (palms or quarters), and sixths (*jemes* or *compases,* that is, the maximum stretch between the thumb and index finger). The foot, in turn, was divided into 12 inches (the doubled-over thumb) and into 24 fingers.

Solid or liquid materials also had measures which, when they were not anthropometric, derived from very ancient everyday usages, including culinary ones: the fingernail, pinch, handful, *almuerza* (the hollow of the joined hands), cup, pouch, sack, *arroba* (25 pounds), "quintal" (4 *arrobas*), glass, wine jar, bucket, shovelful and other even more imprecise ones, like the load, cartful, cart load (of sand, lime or stone) and many more. Weights included, in addition to the *arroba* and *quintal,* the *tonelada,* that is, the traditional Spanish (not current English) ton, derived from a tun - (*tonel*) or barrelful of oil, to say nothing of all the imaginable versions of the Roman pound and Arab ounce.

In previous studies[28], the author of this essay sketched out a theory that domestic building -whether urban or rural - is a genre dominated and characterized by technological considerations, not formal, stylistic or simply aesthetic ones. This theory is confirmed by the proven fact of an utter absence, in the Neo-Granadian colonial period, of architects, or the equivalent of what was then understood as a professional with an academic training. The hacienda houses were the work of building masters, qualified craftsmen, masons and carpenters, not of designers who had been trained in a formal school. Each of these builders had memorized his own, personal "Building Manual", having learned the subject as an apprentice on building sites during his adolescence and/or through the verbal instruction of his parents and relatives. Why recur to books on the theory or history of architecture when the mason usually didn't know how to read? What was the point of an architect's supervision or interference in an architectural genre which, by definition, lacked a superimposed stylistic dimension, much less any kind of evolution in the ordering and treatment of the constructed forms or the interior space? If landlords and builders completely agreed on what a house should be, how it should be built and what it would look like once it was finished, there was no need for a prior design stage, given that an architect could only suggest decorative touches or the luxurious use of materials that implied highly costly cosmetic changes and a certain clash with the aesthetics of spatial codes that were the product of purely utilitarian considerations.

In *Casa Colonial* we referred to the only exception, in Nueva Granada, to the general rule of an architecture of builders rather than architects: the hacienda house of *Aposentos* in Simijaca (Cundinamarca): *"...in what is now Colombian territory there is only one documented case of a hacienda house attributable to an 'architect', Friar Domingo de Petrés, a Capuchin priest from the village of that name in the province of Valencia (Spain), who was presumably responsible for the design and building (or remodelling) of 'Aposentos'...Arriving at Nueva Granada in the second half of the eighteenth century, Friar Domingo had had some academic training in architecture and he continued to broaden his knowledge, in a self-taught way, after entering the Capuchin Order, but he was not an architect who had been officially licensed by the Spanish authorities."*

What continues to be a mystery are the reasons why the owners of *Aposentos* recurred to a person dedicated to religious architecture for a project so alien to his experience, that is, a rural edification. The case of *Aposentos* becomes even more interesting when we consider that, very close to the house which Petrés is thought to have built, there is another which shows signs of being much older than the former and was possibly the original house on the site. If this were so, Petrés might have been commissioned for snobbish motives, that is, to create a distinct architectural reality that was obviously intended to establish an aesthetic superiority to the other hacienda houses in the region.[29]

It is evident that, no matter what the period, an architect would be superfluous in a constructional genre like the hacienda houses, where the entire spatial organization and the technical decisions were predetermined in accordance with a tacit and traditional agreement between the owners and builders. In principle, the basic intention was elemental: to construct some sections of spaces that were generic but versatile and lent themselves to being subdivided (by preference, transversely) into as many compartments as the utilitarian needs of the moment dictated and were able to provide lodgings for the proprietors, including, on occasions, a hall or dining room, servants' quarters, storerooms, closets, kitchens, stables, barns or granaries, all arranged as rationally as possible. This rationality would be assigned, as mentioned above, to two kinds of methods of ordering: one, around a central open space, and the other, a compact one, with perimetral galleries around a built nucleus. When there was a need for them, buildings with specialized functions that were detached from but based on the structural principles of the main house might arise, such as housing for slaves or farmhands, stables, granaries and outbuildings for milling sugar cane or timber. Some of these outbuildings might, in the same way, be adjusted to the volumes of the main house. All of this would be governed by a technological concept that simplified

Opposite page, above, The house of an olive-plantation hacienda in the sierra of Santa Lucía, province of Cádiz, Andalusia, built at the beginning of the eighteenth century.

Opposite page, below, The hacienda house of *Aposentos,* Simijaca, Cundinamarca, built during the last years of the eighteenth century. Note the volumetric similarity and the symmetry in the façades. The use of corner towers as a stylistic accent is the same in both cases and forms part of a long tradition in Andalusia and Levant that originates in castle and citadel architecture.

everying: without roofs there are no spaces and without spaces a house does not exist. In this way the *a priori* determination of the dimensions of the spaces to be obtained through estimates of the length and cross-section of the wood needed for the frameworks of the roof that would enclose these spaces is primordial. This would not be a matter of architectural blueprints or designs but would involve, instead, a calculation of the quantity of building materials, made on the basis of a description of the proposed house which might be only a verbal one.

The length, width and number of the constituent sections of the house, together with the height of the roof, provided sufficient data to begin the work. On this basis you could determine how much stone from the river was needed for the foundations; how much lime and sand would be required for the mortar and plaster; how many mud-bricks for the walls; how many trees would have to be cut down to produce wooden columns, lintels, plinths, tie-beams, rafters, ridgepoles, as well as the planking and frames for doors and windows. And, finally, the decisive and costly question of how many thousands of ceramic roof tiles would have to be made to replace the straw thatched roof with something more durable and more reminiscent of Andalusian or Castilian building traditions. Then came the complex, lengthy and difficult part. To obtain the right wood you had to know when and how to cut species of trees which had not grown in accordance with the biological calendar of the annual seasons. It was vital to know or remember a great deal about European tree species in order to find, in the midst of the unprecedented natural wealth of the landscapes of Nueva Granada, the exact or loose equivalents of the woods that were suitable for structural work or decorative carving.

In addition to what the Spanish settlers, through technical analogies or ancient remembrances, would call oaks, cedars, walnuts or chestnuts, a surprising abundance of native species were also employed (ceibas, carretos, abarco, amarillo, morado or amaranto, comino, granadillo, canelo, zapatero or boxwood, caobas, sapanes, guayacanes, mangles, etc...). By virtue of their structural importance, such woods were the dominant technical note in the construction of rural houses. The factor which established the most convenient and practical dimensions for the interior spaces of such houses was not some abstract caprice but the length and resistance to bending of the timbers at hand, that is, the spans or intervals which might be bridged by these, either from wall to wall or from a wall to the columns of a patio. The size of the timbers would, in turn, be determined by the age, height, trunk diameter and biological characteristics of the trees to be found at a reasonable distance from the place where the house was to be built. This is the reason for the inevitable and universal formal and atmospheric similarity of the totality of the interior spaces of the Neo-Granadian hacienda houses, no matter what the region or climte. What would come to establish infinite variants within this presumed uniformity would be the relation between the landscape and the exterior volumetry of the house and the usages that would take place within it, but not the constant or aleatory application of a given principle of spatial ordering or a compositive, that is, design formula.

If the consecrated tradition of wooden frames placed on bearing walls was never broken or subjected to any basic alteration, how could an architectural evolution of the genre ever have occurred? It should be noted that groups of builders who knew of (or practiced) the techniques of flat or terraced roofs never reached Nueva Granada, except for a very few who built urban and hacienda houses in and around Santa Marta. The most celebrated of these few exceptions is the *Quinta* (that is, the chocolate bean-mill estate) of San Pedro Alejandrino, whose main house, famous for being the place where the Liberator, Simón Bolívar died, is completely covered by flat roofs. In the rest of Nueva Granada the roofs of rural buildings were unusual or orthodox but always highly economical or rustic versions of the wooden frameworks of Islamic (Andalusian) origin known as "par e hilera" and "par y nudillo".[30] These techniques for building a roof with a minimum of timber and dead weight that would also be structurally independent of the support walls completely dominated the construction of all architectural genres until the end of the Colony – an unusual case of technological unanimity and uniformity. Unfortunately very few colonial documents that directly refer to the technical aspects of building hacienda houses have been conserved, while there is an abundance of those which refer to land deeds and lawsuits over land deals and ownership, as well as to disputes over sources of water and boundaries. Luckily there does exist one such document, already cited in several books and originally published by Camilo Pardo Umaña in *Haciendas de la Sabana de Bogotá*.[31] It deals with an account of expenses made, in 1770, by the builder (*alarife*) Francisco Javier Lozano for work on the hacienda house *La Conejera* (Suba, Cundinamarca). According to Lozano, the house had, at that time, "6 stretches of 6 and 1/2 *varas* [about 5.44 metres] width". Assuming a width of 2/3 of a *vara* for the bearing walls, you would obtain a space or distance between the walls of 5 and 1/4 *varas* [about 4.33 metres], which would be a normal or average space between such walls for that period and for the savannah of Santa Fe. Assuming an average length of 14 *varas* for each section, the building would be a house of some 546 square *varas,* that is, about 426 square metres. This seems to fit the case of one of the buildings there, thought to be oldest of those which came to be built in *La Conejera*. The accounts, rendered in *patacones* of eight tenths – the gold coins that were the "official" legal tender of the period – are as follows:

20,009 cart loads of split stone at 2 reales [coins] each	*500*
124 tapias [modules or sections of dried mud wall] *at 4 reales each*	*62*
20,000 adobe bricks at 3 pesos per thousand	*60*
2,000 bricks (baked) at 12 pesos per thousand	*24*
2 columns of stone with their bases and capitals	*150*
1 stone portal for the oratory	*28*
12 varas of stone for masonry at 2 pesos the vara	*24*
9,000 roof tiles at 13 pesos per thousand	*117*
1 flight of stairs	*10*
12 varas of medium stone for masonry at a peso each	*12*
Labour costs and wage of the building officer [the assistant mason]	*800*
Total	*1,787*

The above shows that we are talking about an "average" rural building for the period, one without luxuries but not very inexpensive either. Note how labour costs represent nearly 45% of the whole, a level which can also be found in other documents that refer to the building of colonial churches and convents. If building materials were cheap, manual construction work had a very high price in the context of the colonial economy. The two columns and what must have been a small portal for the oratory, together with the masonry stone and the split stone for the foundation, came to 714 *patacones,* no less than approximately 40% of the total expenses. In fact, stone was always the most costly material in every colonial building, no matter whether it was simply "split" or carved, with a fairly mediocre ability, into columns, bases or capitals. This more than explains the proverbial scarcity or absence of stone parts, not only in hacienda houses but also in their urban equivalents and the preference for the use of wooden columns and lintels instead of arcades of stone columns and brick arches.

It was obvious that the quarrying and transport of stone were more troublesome and implied bigger costs than the felling and haulage of timber, a situation which still applies today. Thus, a stone portal of the size and sculptural quality of the main entrance of the house of *Aposentos,* Simijaca, which perhaps was designed by Friar Domingo de Petrés, may be unique in the rural architecture of Nueva Granada. It is best to distrust the authenticity of others of the same genre that were acquired from urban demolitions and taken to the countryside or ordered from a present-day stoneworker to "improve" the entrance to a family farm.

Some idea of the magnitude of the absurdities committed, in more than one colonial hacienda house, by 20th century "restorers" and decorators who have aimed for a pretentious elegance by replacing the humble galleries of pillars and lintels with arcades supported by columns acquired from the demolitions of cloisters and urban houses may be gathered from the worst example, in the country, of this class of outlandish architecture– the old house of the hacienda *El Salitre,* in Paipa

(Boyacá), whose innumerable stone columns were placed there in the nineteen-fifties and largely came from demolished houses and convent cloisters in Tunja and other parts of the country.

On occasions the recent insertion of stone columns into a modest rural house looks clearly surrealistic, as in the case of *El Noviciado,* a farm in Cota (Capital District) that resulted from the subdivision, in the 18th century, of the hacienda *Buenavista,* whose rustic wooden columns in its only ground floor gallery were replaced, in the fifties, by the big stone columns visible today, which came from the demolition of the old part of the priest's house adjacent to the church of the neighbouring town of Chía – a gift from an ex-president of Colombia, who apparently "didn't have anywhere to put them". Given that the rest of the house didn't benefit from some other architectural gifts, the contrast between the stone columns and the humble components of the eastern façade of the building is, to say the least, disconcerting.

The preference, shown in the above document, for different kinds of earth bricks is easily understood. The cost of brick, including its cumbersome transport, was four times that of earthern blocks, given that it implied the use of pure clay, compressed and moulded by hand and then baked with a great expenditure of wood or coal. The two main kinds were adobe, mud mixed with straw that was kneaded and dried in the open air and *tapia*, a mixture of earth, sand and pulverized stone, with occasional additions of lime, that was rammed down into wooden frames, after having been "steeped" for a while (a technology that varied from region to region, according to the availability of materials). Neither required the complicated process of firing used for brick and roof tile, so that the relative cost structures were greatly different. To raise walls of adobe stuck together with wet earth or tamp down *tapias* did not require the use of skilled labour, but brick and mortar masonry work and roof tile laying needed additional experience and knowledge, which made them more costly. For this reason the purchase of one clay brick for every ten adobe blocks in the construction of *La Conejera* leads us to assume that brick was going to be employed in one of two possible ways. As a floor finishing in the reception halls or in the family members' bedrooms, since the usual flooring in the rest of the house was tamped down earth covered with floor mats made of handwoven plant or palm fibres. Or for horizontal rows, at regular intervals, in the adobe walls, to shore them up and make them more resistant to the frequent vertical fractures. These supporting rows, called *verdugadas* in Spanish or *rafas* in Arabic, are another technical tradition with joint Roman and Islamic origins.

Indigenous techniques, although largely replaced by European traditions that were executed in native materials, were not entirely abandoned. It is clear that indigenous huts were used as temporary shelters at the beginning of the colonial period, but only when such dwellings happened to be found in places where the Spanish were going to establish permanent settlements: their employment was also limited by the official prohibition (rarely respected) against "bothering" the naive communities. Spanish or *mestizo* builders never had the slightest interest in (or worthwhile motives for using) anything other than European constructional techniques, adapted to the Neo-Granadian environment, for the very simple reason that their ideas about existential space or their perception and ordering of artificial spaces were totally alien to those of the indigenous groups. Some indigenous technologies (that is, another order of ideas), like that of *bahareque* (mud laid on a framework of interwoven cane) were easily integrated into Spanish construction techniques and continued to be used throughout the colonial period, above all to build partition walls and lean-tos. Others, like the use of *cañabrava* and *chusque* (varieties of rushes or reeds with a cane-like stem found, respectively, in swamps and prairies), tied with hemp or plant fibre, replaced (demonstrating a praiseworthy sense of the practical) the costly and difficult-to-produce planks which formed the support base for tile roofs in Andalusian construction. The regional and *mestizo* use of the "cake" of clay over a base of *chusque* was obviously less durable, heavier, less costly and more difficult to maintain in a good state than the Spanish

Opposite page,
La Conejera, Suba, Capital District.
The transformation of the colonial house during the nineteenth and twentieth centuries has been complete. Its current appearance completely belongs to the modern period.

plankwork, but the non-evolving technology of the colonial period never went beyond it. In all this we should bear in mind the very rudimentary manufacture and use of the tools and machines used for carpentry in Nueva Granada, one indication of which is that the work of removing bark, "squaring off" the wood and smoothing down wooden beams and columns was only done with hatchets and carpenter's adzes. The rough finish that was thus obtained was noticeable, above all, in rural construction, where, more often than not, no one was willing to pay for highly skilled work.

One of the favourite resorts of present-day interior decorators consists of using decorative wooden beams that have been carefully cut by machine but are later knocked about and mistreated to obtain a false (or painfully cosmetic) version of the rusticity which, in the original building of hacienda houses, was about the most polished that could be hoped for. Of course, the use of a basic carpenter's plane, or gouges and chisels, although familiar to colonial craftsmen, was reserved for certain works of carving in urban buildings of a certain luxury status or for furniture-making. The inclusion of beautiful columns of turned or sculpted wood in hacienda houses (such as that of *La Concepción de Amaime*, El Cerrito, Cauca Valley and some others in the same region) was highly exceptional.

The document about the building of *La Conejera,* in the savannah of Bogotá, reflects a notable aspect of colonial building technology: the organization of the workforce. Neither a building master (*maestro de obra*) nor a foreman (*contramaestro*) was employed to oversee the construction of that hacienda house. The person who did the work was a simple builder (*alarife*), whose rank would correspond to that of a mason in the republican or modern period in Colombia. Like the common labourer, the *alarife* had callused hands and had to climb the scaffolding and stick adobe together: he did not have the authority to simply give orders and wait for them to be carried out. According to his list of expenses, he worked with a building "officer", that is, an assistant, whose title and rank are exactly the same today in Colombia, which is also true of the anonymous helpers or apprentices who are responsible for hauling materials and other "heavy" jobs. The inventory of *alarife* Lozano only refers to the fabrication of walls and floors. According to the division of labour in that period, the roof frame would have been the responsibility of a cabinet carpenter (*carpintero de lo blanco*), that is, one with a workshop and more specialized tools who was responsible for the structural components in wood. Then there would come the tiler (*entejador*), who would finish the roof and the ordinary carpenter, in charge of making the doors and windows. But if the work was being done in a rural place that was remote and hard to get to it was not unusual for a single builder to do the work both of mason and carpenter, thus becoming the forerunner of the *maestro todero* (jack of all trades) of nineteenth and twentieth century Colombia, a species that is practically extinct nowadays.

Today hacienda houses, with only a few exceptions, summon up the idea of a simple, rustic atmosphere that has paradoxically become part of their charm. We speak today of the enchantment of thick walls that, victims of their own dead weight, become cracked, fractured or twisted and are picturesquely patched or bent over; of poorly cut and assembled timbers; of roofs that are poetically bent because of the weight of their tiles and clay-and-cane bases and have never been repaired in time. In short, the charming world of an architecture created by those who didn't know any better nor were able (or wished) to do anything else. Which is also the world of those who unfortunately forget that the sorry idea of "the picturesque" is only suitable for mediocre painters and dilettantes. It is ironic that technological backwardness should, with the passage of time, be converted from a deficiency into a virtue. Nowadays, the poorer the quality of its construction the more beautiful and evocative becomes a country house in the eyes of those who gaze at but do not have to live or work in it. According to the Colombian writer, Eduardo Caballero Calderón, in *Diario de Tipacoque*:[32.] "...Country houses are beings that grow old, become wrinkled and wind up laying themselves down to die by the roadsides, but they last for centuries, strong and vigorous ...conserving the sonorous echo of their rooms, the resonance of their winding corridors, the living epidermis of their walls, which are cracked and tanned by an implacable sun…".

Whoever undertook the difficult endeavour of building a hacienda house in colonial Nueva Granada, while he might have known nothing of the artistic theories or cultural meanings behind his work, did nevertheless intuit a great deal about its aesthetic and atmospheric qualities.

This is the only way we can explain why certain hacienda houses, in regions so distant from one another as the east of Boyacá and the south of Cauca, show a surprising purity of form and proportion within the rigid parameters of colonial technologies. The exquisite proportions seen in the structural modulation of columns and lintels in the galleries of the ground and upper floors of the house of *Calibio* (Popayán) – one of the best examples of this – are based on a simple technical rule: the square modules of the façade are the result of an employment of vertical and horizontal pieces of wood with the same length and imply the least possible waste of an expensive and difficult-to-obtain material. The squares, in turn, are figures which allow for multiple geometrical combinations that are almost always harmonious and beautiful, in which variations introduced for technical reasons do not alter their basic nature but assume remarkable aesthetic qualities. It is possible that, in the infrequent cases when landowners and builders agreed, more or less instinctively, that the width, length and height of a drawing room should form a dimensional combination that pleased the eye, thus brushing against the idea of the "Golden Section", these aesthetic qualities were not entirely accidental or fortuitous. Among the many builders of hacienda houses, whether experienced or not, there must have been some who had a real artistic talent, even if it was a latent, not a cultivated one. What is more, as bricklayers and musicians know, he who has a complete technical mastery of a calling will almost always become a virtuoso.

Colonial constructional technology was not a single body of ideas or knowledge. It showed considerable qualitative variations in different regions of Nueva Granada. The aptitudes, training and resources of the craftsmen who engaged in construction played a part in this: they ranged from the inspired to the extremely sloppy. Beyond this, the availability and rational use of local materials were decisive in an ambit that was predominantly technical. We might conclude that rural building in Boyacá in the seventeenth and eighteenth centuries was qualitatively better than in Cauca, with better-built walls and better-treated and assembled timbers in the roof frames, but this would only be a simple confirmation of accidental circumstances. Namely, that the clay for roof tiles and bricks was of better quality in the high plains region of Cundinamarca and Boyacá and there was a greater availability of skilled masons and carpenters. The obvious technical inferiority of colonial roof frames in the region around Popayán, compared to their equivalents around Santa Fe, was due to an exaggerated tendency to simplify and lighten, perhaps in order to reduce costs and not necessarily because of technical ignorance. The suppression of structural components and the building of roof bases in materials, like cane, which are fragile and short-lived, was always a bad habit that led the resulting houses into a premature obsolescence or decay.

"In what is now Colombian territory there is only one documented case of a hacienda house attributable to an 'architect', Friar Domingo de Petrés, a Capuchin priest… who was presumably responsible for the design and building (or remodelling) of the house of Aposentos…".
G. Téllez
Casa Colonial.

Above and right, *Aposentos,* Simijaca, Cundinamarca.
The house combines a sure intuitive sense of place, which is apparent in its location in a dip in the surrounding hills, with the symmetrical elegance of its volumetry and main façade, which has an obvious academic origin. The result is the most untypical but most architecturally interesting hacienda house of the colonial period in Nueva Granada, even though it was built at the end of that epoch. The enclosure walls of the fields near the house extend its spatial command of the place where it is situated.

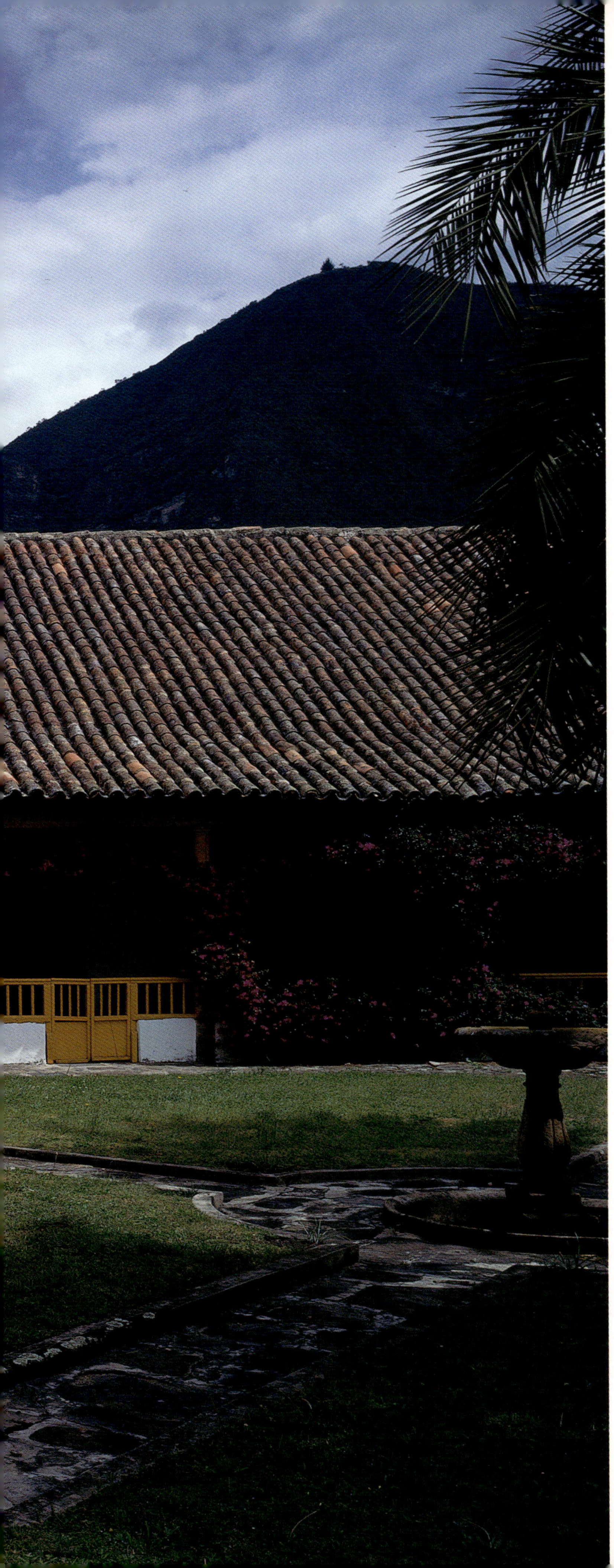

Left and above, *Aposentos,* Simijaca, Boyacá.
The generous proportions of the main patio of *Aposentos* reflect its theoretical origin. Note the contrast between this patio and that of the house of *Gotua,* Iza, Boyacá. Both adhere to the same concept of spatial ordering and were built with identical materials and constructional methods, but the respective atmospheric results are very different from one another. One was made simply to be lived in, the other to contemplate its formal elegance.
The inevitable alterations done to the house in the republican epoch (flat ceilings, veneering and moulding of wooden columns, the placing of railings over the benches bordering the patio, etc.) marginally affected the atmospheric quality of the spaces. Note how the dominant presence of the artisanal tiled roofs seems to belong to a conceptual world that is very different from the modular orthodoxy of the colonnades around the patio or the ordering of the main façade of the house, as though the intention had been to roof the house with flat terraces in the manner of the Spanish province of Valencia, birthplace of its designer, Friar Domingo de Petrés.

Left and above, *Yerbabuena,* La Caro, Chía, Cundinamarca. It appears that the original sections of the house of *Yerbabuena,* which form an "L" on two sides of the interior patio, were built at the end of the eighteenth century, when the pastures of *Hatogrande* were dismembered and several haciendas arose from this subdivision. In the first half of the nineteenth century the house was gradually enlarged and reformed to reach an unusually large size compared to its counterparts on the savannah of Bogotá. In the second half of the nineteenth century the republicanization of *Yerbabuena* gave it the architectural flavour which is seen in the most "modern" part of the house (above), which underwent several major alterations. This included the closures, in "French-style" glasswork, of the galleries and the pseudo-Sevillan iron grilles. The "box"-type eaves, usual in the hacienda houses of Cauca but surprising in the middle of the savannah of Bogotá, that are seen here in the façades and towards the interior patio were imposed in the nineteen-forties.

Above and right, *El Colegio,* Madrid, Cundinamarca.
It owes its name to the acquisition of the surrounding lands by the Royal Seminary College, an entity which belonged to (and was administered by) the Jesuits. Before passing through the hands of the Jesuits it formed part of the vast properties of the *Dehesa de Bogotá* estate. Thanks to some legal tricks by the Jesuits, the hacienda was not expropriated by the colonial government when the Jesuits were expelled from Nueva Granada in 1767. By that time the house had come to have the volumetry and size that we still see today. It is one of the few Neo- Granadian hacienda houses that we can definitely say were built at the beginning of the seventeenth century, that is, at least parts of it, including, perhaps, two sections of the ground floor.
Its conservation by succesive private landowners saved it from the abandonment and official vandalism that other rural properties of the Jesuits were subjected to.

Left and following page, *El Colegio,* Madrid, Cundinamarca. The transformation of the interior of the house of *El Colegio* during the end of the nineteenth and beginning of the twentieth centuries was very extensive. It included the repairing of the two flights of the large stairway and the space it occupies, as well as the hanging of wallpaper, a decorative resource usually applied to urban residences, and flat ceilings common to almost all the hacienda houses of the savannah of Bogotá. In many cases the interior subdivisions increased the number of bedrooms while reducing the size of the original ones. This process extended from the cities to the countryside, where the amplitude of habitable spaces does not depend on the limited size of urban building lots.

Following pages, *El Colegio* has one of the best- proportioned and elegantly modulated two-storey patios of the hacienda houses of the savannah of Bogotá, despite the presence in it of a well mouth and stone fountain whose size would be more appropriate to an urban plaza.

"There suddenly appeared, on a hill, the poetical hacienda of *Cincha,* from which you have a very beautiful view. To the left, the curious high plateau called La Mesa, which rises above the hotlands. To the right, *Canoas,* where the proud deer run, which are abundant there ". Miguel Cané. *En viaje* ("On a journey"). 1913.

Above, right and opposite page, *Cincha,* El Charquito, Cundinamarca. Built towards the end of the eighteenth century as a result of the subdivision of the hacienda of *Tequendama,* it conserves, its discreet and evocative main patio, whose setting includes such original elements as the supports and plinths, as well as the structure of wooden pillars and lintels in its perimetral galleries. *Cincha* was not subjected to a republicanization or an intensive modern alteration. No ceiling alters the interior spaces nor prevents one from appreciating the roof frames, which are built in a magnificent Santa Fe (Bogotá) version of the Andalusian "par y nudillo" technique This explains the importance of *Cincha* as an invaluable example of what a modest hacienda house was like during the late- colonial period.

Opposite page and right, *Hatoviejo,* Yotoco, Cauca Valley. Like *La Sierra* and other Cauca Valley hacienda houses, its spectacular siting is its principal architectural merit. Built starting from the last years of the seventeenth century, it came to have a spatial form in the shape of a "U", possibly because of successive extensions of the first sections. That is, it is made up of three sides of a patio which never acquired the fourth and last one. The resulting concavity acts as a focal point for the panoramic view of the valley.

Left, The house of *Hatoviejo,* like *La Concepción de Amaime,* had a separate chapel, whose location is curiously deflected from the house. It now has a modern portal.

Above, The volumetry achieved by the house of *Hatoviejo* at the end of the eighteenth century is, with the exception of some inevitable modern additions and reforms, the one which we may observe today.

Previous pages, left and right, *Fagua,* Cajicá, Cundinamarca. It has an irregular volumetry and spatial organization compared to the patterns usually seen in the haciendas of the savannah of Bogotá. It combines one and two stories around several successive open stories, which is an indication of a large and complex growth by stages, as the house passed through the hands -and the reforms-of many owners and was subjected to a number of architectural plans that were never completely realized. This gives a certain flavour and informality to the house. Starting from the last decades of the sixteenth century what must have been a temporary shelter was erected on the lands which would later be grouped together under the name of *Fagua*. The first "solid" sections of the original house seem to date from the first third of the seventeenth century and the present form and size of the house were reached during the first half of the eighteenth century. Despite periods of abandonment, the house shows a remarkable degree of conservation, especially its exceptionally well-maintained portals and surrounding walls.

"It is evident that an architect would be superfluous in a constructional genre like the hacienda houses, where the entire spatial organization and the technical decisions were predetermined in accordance with a tacit agreement between the owners and builders." (From the text of this book).

Left and above, *Fusca,* Torca, Cundinamarca. If the house of *Fagua* is an irregular spatial organization, which was arrived at as the result of a fragmented process of reforms and alterations, *Fusca,* by contrast, displays the spatial orthodoxy characteristic of an architecture that was built all at once, with a single guiding principle. *Fusca* dates from the years 1778-1780, which makes it a comparatively late example, and was designed with the idea of precisely adapting it to the dip in the terrain and later, with a certain rigour, of arranging it around a central patio, although it also has a large gallery, in the manner of a reception hall, which looks out on the view of the adjacent savannah. The latter does not contradict the central patio. On the contrary, it includes the patio in the logical and beautiful sequence that takes place as one wanders through the house, which is approached from the back, not the front: one then walks round the patio, passing through the main reception rooms to come out, at last, on the gallery-balcony of the front.

Left and above, *Fusca,* Torca, Cundinamarca.

Left, The passage of the nineteenth century is evidence by the ceilings, doors and windows of that epoch. What would have been an expressive and picturesque spatiality, with the usual variegated complex of roof frames of round rafters and a base of cane and clay, is hidden today by ceilings decorated with some Italian-style touches of plasterwork that are strangely incongruous with the colonial origin of the house.

Above, An example of how history plays a role in hacienda houses: the bedroom where the Liberator Simón Bolívar spent the night. Bolívar had a profound liking for (and understanding of) hacienda houses, which shaped his life from his childhood to his death.

Left and above,
La Esmeralda, Tabio, Cundinamarca.
Part of the rural buildings created at the end of the colonial period, when the dismembering of the great estates of the savannah of Bogotá intensified. At some stage the enormous property of *El Novillero* was divided into eleven parts, one of which became *La Esmeralda*. Although colonial hacienda houses date from different periods and have widely varying sizes, the charm of their tiled roofs and the delicate relation between the house and place are features. In *La Esmeralda* the imaginary prism of space which the house is inscribed in is almost tangible and the house itself looks as natural as the trees which surround it.

Left and above, *El Portezuelo,* Ubaté, Cundinamarca. The favourite location for the houses of this region is at the foot of a hill or mountain, where springs of water or streams are more likely to be found and the house may be sheltered from the evening wind or dawn cold. In Boyacá and the north of Cundinamarca one commonly finds a house with a compact floor plan, provided with a separated but adjacent hut or pavilion for the foreman and his family, storerooms and stables. *El Portezuelo* seems to have been consolidated at the end of the eighteenth century and the present form of the house may date from the same period.

Left and above, *Bomboná,* Nariño. It is situated in a mountainous region where haciendas were scarce but covered remarkably large areas. Few large estates in the south of the country came to have houses of architectural importance and few of these survived into this century: Bomboná is so exceptional. Its construction must have required a great number of journeys to the nearest source of roof-tile clay, good adobe and, especially, suitable lime, all of which lay leagues away. And add sthe devastation of an enormous area of forest for the timbers that went into the thick columns, lintels and plinths that make up the tremendous structure of roofs and galleries of the house. When one sees a hacienda house in a remote part of the countryside one tends to forget the impressive technological endeavour that lay behind its construction and played a decisive role in its architectural character.

Left, The unusual lookout tower which breaks the volumetric unity of the tiled roofs. Very little is needed to break up the continuity and formal harmony of colonial tiled roofs. This tower, which was indispensable for keeping an eye on the surrounding area in a region with a long and troubled history is not an original feature of the colonial house built in the middle of the eighteenth century but was added in the first half of the nineteenth century.

Right, *Papare,* Ciénaga, Magdalena. Situated at the other end of Colombia, the house of the sugar-mill hacienda of *Papare,* Magdalena, is an architectural example that is as exceptional and isolated as *Bomboná.* The building techniques found in the urban houses of Santa Marta and Ciénaga were obviously utilized in *Papare,* including roofs in flat terraces with attics, which is similar to those found in *San Pedro Alejandrino* in the same region. These two examples from the Caribbean coastal region are an exception to the omnipresence of tiled roofs in the rural houses illustrated in this book. *Papare* also has an undeniable resemblance to an urban house, due to the use of projecting perimetral balconies instead of broad galleries built on wooden columns or pillars. In the separation between the balcony floor and the small roof over the windows they resemble those which appeared throughout the cities of the Caribbean coast during the nineteenth century.

Myths and realities

"The house is the conscience of the countryside.
The countryside turns to look at it from every angle
and sees itself in the house as though it were seeing a
fruit of the earth's innermost necessity."
EDUARDO CABALLERO CALDERON.
Diario de Tipacoque

As stated in previous chapters, Andalusia marked the last stage in the process of architectural transculturation which brought urban and rural houses to the New World, but this does not mean that these houses were of an exclusively Andalusian nature or that they were a simple copy of Mediterranean concepts or traditions. In a manner of speaking the Neo-Granadian hacienda house is, in its own way, original, having acquired such a character through an almost biological process of adaptation to the environment of the diverse geographical regions where it appeared. In *Casa Colonial* we showed how what is American or Latin American about the colonial house is precisely that it was built in the New World, although it might have been conceived, centuries before, in the Middle East and parts of southern Europe. It is essential to bear in mind that the materialization of a constructed form in a specific place modifies, up to a point, the idea which initially created it, but this alteration is often a matter of degree, not of kind. The building of a country house depends almost completely upon what the countryside provides for that task and so, the place and the house wind up belonging to one another in a physical and metaphysical sense. A Neo-Granadian hacienda house is such, for having been built on American soil, although with European ideas. Its *origin,* that is, its originality as an architectural species is to be found in a geographical, not an ideological, mode. Building a farmhouse on the outskirts of Carmona, in the province of Seville, is by no means the same as building a hacienda at a certain distance from Popayán, although the governing principles for the interior and exterior spaces may be generically the same and one may find many similarities between their respective landscapes. The elegant pretensions of the Andalusian farmhouse of *La Baldía,* for example, seem to belong to a world that is formally very distant from the modesty and reticence of the house of the Cauca hacienda *Antón Moreno,* although, conceptually, they show certain analogies. One continues to be an exclusive product of Andalusia, while the other was only created for a site in Nueva Granada. Rural architecture is, precisely, the genre in which material circumstances - the place - dominate ideas, while in cities and towns concepts of ordering, especially if they have an academic origin, are predominant.

According to the Spanish historian Mario Sartor, in his essay *La vivienda mediterránea y la tipología de la casa colonial americana,*[33.] the rural house is an architectural *whole* that does not lend itself, as does the *domus,* that is, the urban house, to a fixed typology. This assertion is typical of this historian, who does not find, in rural buildings, the formal analogies necessary to classify their anarchic wealth of variations within typological groups. He thus resolves the question by alluding to the abundance of "parallel models" in the architecture of the countryside, whose parallelism necessarily has to occur with respect to their urban counterparts or else among rural buildings of different geographical regions. Situating the haciendas of Nueva Granada in the context of the American continent, one sees how there are inevitable analogies in socio-economic origin, land holding and the modalities of land exploitation, but it is also possible to establish significant differences in the historical processes behind the forming of the productive units and even in the "parallel models" of architecture for rural buildings in different regions.

The hacienda houses of Nueva Granada lack the variety and size of outbuildings that are to be seen in the Mexican haciendas, for example. While, in Nueva Granada, there only existed two basic kinds of haciendas, one for cattle-rearing and the other for farming, and among the latter two sub-types, the sugar mill hacienda (as in the Cauca Valley) or timber mill hacienda (as in Boyacá), on the one hand, and the hacienda lacking such elements of production, on the other, in Nueva España (Mexico) there came to exist not only grain-producing or cattle-rearing fincas but also ones that were

Opposite page, *Coconuco,* Cauca.

dedicated to such activities as mining, the specialized cultivation of cotton or agave fibre (*fique* or *henequén*), the production of such liquors as *pulque* and *mezcal,* special forestry products (precious or exotic timbers), etc. Each of these activities created distinct architectural types of housing complemented by the different structures that were destined to lodge ever more specialized functions. In the case of Nueva Granada, only the sugar mills known, respectively, as *trapiches* (employing animal power) or *ingenios* (employing hydraulic power) came to require buildings that were separated and radically different from those of the landlords' houses, as may be observed in *San Pedro Alejandrino* (Santa Marta) or *Piedechinche* (Cauca Valley).

The stupendous formal and technical variety found in Mexican silos, barns, reservoirs, aqueducts, waterwheels, ovens with enormous chimneys and watchtowers was never seen in Nueva Granada, for the socio-economic reasons already explained. The opposing technical tendencies on the Latin American continent were, on the one hand, the employment of generic, not specialized, spaces for the functions of housing and work, usual in Nueva Granada; and, on the other, a growing specialization and sophistication in architectural spaces that functioned around their eventual use, as in fact happened in Mexico, Cuba and Brazil. When, in the second half of the twentieth century, there finally and inevitably arose good-sized industries in the immediate vecinity of the hacienda houses of the colonial period, as was the case with the dairy industry beside the hacienda *Fagua* (Cundinamarca) or a number of industrialized sugar mills in the Cauca Valley, it occurred with a delay of nearly two centuries compared to Mexico.

It is possible to establish, in a very broad way, that the hacienda house, in Nueva Granada, is made up of the following constituent elements: The house of the landlords (*los señores*), including bedrooms, one or more halls acting as dining and/or reception areas, and with accessory outbuildings such as the open-air baths (*chorros*) of the haciendas of Cauca, or the free-standing or interior chapels or oratories, like those found in *Calibío* and *Antón Moreno* on the outskirts of Popayán. An extension of the above would be the collection of what are today called service areas, which were usually grouped around a separate free space and included kitchens, pantries, laundries and complementary buildings. The kitchen might form part of the house, generally being arranged in the Andalusian fashion, that is, with the whole roof of the space devoted to cooking and food-preparation forming a chimney, but for fear of fires, which were frequent, it might also have been located in a separate hut or shack, built with indigenous techniques and materials. In the haciendas in the western part of Nueva Granada, lodgings for permanent workers (*peonías,* from the Spanish *peón*) or slaves had a certain importance and were combined, at times, with granaries or storerooms, or stables and stockyards. These outbuildings were invariably built at a distance from the house of *los señores* but, in the houses of smaller farms in Cundinamarca and Boyacá, the servants' quarters were frequently an extension of the kitchen and pantries. It is worth noting that the stables found in the hacienda houses of the high plain of Cundinamarca and Boyacá were frequently built as an extension of what would eventually become the real entrance to the house, on the opposite side of what would have been the house's "main front".

We mentioned above how the hacienda and farm houses of Nueva Granada showed variants of spatial organization derived from only two basic schemes of Mediterranean origin: The first, organized perimetrally around a central open space and formed by sections of closed volumes. Between these and the central patio (or courtyard) the galleries or corridors are situated, which are transitional spaces between the completely enclosed and the completely open. This is the plan which offers the greatest possible versatility of use, lending itself, with equal ease, to monastic, educational or institutional functions and for that very reason, utilized without distinction in urban and rural architecture.

The second plan inverts the terms of the organization of spaces: the central part, that is, the patio in the alternative just described, is now occupied by one or two wings of closed and covered spaces, and around these are situated the galleries or corridors, which are now perimetral. Examples of these would be the houses of *Suescún* or *Tipacoque* in Boyacá and *Japio* or *La Concepción de Amaime* in the Cauca Valley. The open space, in this case, is the whole of the natural surroundings of the building. We may suppose that this second scheme is more characteristic of rural architectre, where large free areas are available and the limitations of size found in town and city building sites and lots do not exist. In addition, we may assume that it is easier to achieve a commanding view of the surrounding landscape by using outer galleries (and a lookout balcony, of course). In Nueva Granada there are many hacienda houses which combine the two spatial systems, being organized around a central patio but having, at the same time, perimetral galleries or balconies. This may be found in *Fusca,* Cundinamarca, *Antón Moreno,* Cauca or *El Colegio,* (Cundinamarca). For the rural landlord or builder there was never any contradiction between the inner world of the patio and the extension of the surrounding space in the gallery which ran towards it. The patio was a symbolic piece of the countryside introduced into the house so that one could live in or around it. The surrounding countryside, by contrast, was a spectacle to be seen from a distance and in the round. The house would thus be a reality interposed between everyday existence and a more contemplative reality.

A compositive element as important as the hacienda house itself are its surrounding walls. By means of these the view from the house towards the outside world is organized and measured out, making the latter comprehensible by putting intelligible limits and dimensions on it. The walls inscribed onto a landscape confer an order upon it that is poetical and mathematical at the same time. For the most prosaic and utilitarian reasons, it was also necessary to provide gateways, stockyards, stables and sown fields, thus establishing the double origin of the enclosure walls, perhaps the most fascinating feature of the Neo-Granadian hacienda houses. In *Casa Colonial* we read that:

"The hacienda house is itself and its surrounding walls, its real space and the virtual space determined by the latter ... The boundary walls of stone or humble adobe, scarred by the weather and passage of time, came to delimit, as a first perimeter, the gardens near the house, if there were any, the approach on horseback, the courtyard where the farm equipment and saddles were, the well or the water tank.

"A second perimeter might include the kitchen garden, the chicken coops, the nearest pastures and the orchards. Lastly, a third and distant perimeter, which was barely visible among the bushes and hillsides, would come to delimit the grazing land for the cattle and the bigger cultivated areas. Each of these successive spaces contains all of the previous ones. Each of the previous ones announces and presages the following, larger and more distant one. All start from the centre of the hacienda house."

These considerations about the arrangement of the natural space incorporate one of the most basic elements into the hacienda house: air. Another element, water, also deserves an honourable mention, since it is an essential part of the existence of constructed forms. In the first chapter of this essay we pointed out how life itself and the artificial forms that surround it would be inconceivable without the presence of water. In the arid regions of northern Mexico extraordinary works of hydraulic engineering were needed to bring drinking and irrigation water to many haciendas that were distant from any source or course of water. The cultivable areas of Nueva Granada were more generously supplied with water, so that the enormous efforts, still seen today in what was Nueva España, to give life to the land and those who exploited it were not necessary. The presence of water determined the quality of the lands to be possessed or conquered and colonial lawsuits over the use and sharing out of water courses and sources played a major role in the modest rural history of Nueva Granada.

In the regions of the Andean high plain the building of an aqueduct that was anything more than a succession of grooved stone pieces placed on the terrain, or a simple irrigation channel, that generally emptied into a cistern or well, was rare. Some remains of water supply channels and spouts in the auxiliary outbuildings of the hacienda houses of Cundinamarca and Boyacá

can still be seen, vestiges of the traditional forms of water supply, but all of the stone fountains which have been placed today, for purely aesthetic reasons, in the courtyards of the rural houses of the colonial period are incongrous modern "contributions" (by interior decorators) which falsify the historical testimony which they may provide. They are just another example of the tendency to "elegantize" the modest architecture of the hacienda houses of the colonial period. Although, exceptionally, a cistern or subterranean well might have been placed in the central courtyard of the house (and not outside of it, as would have been more logical and convenient), what you would have seen on top of it was, at most, a modest well mouth for drawing water. It is obvious that what you did find in some interior patios was a simple tank to collect rainwater from the surrounding roof tiles. It is easy to understand the reluctance of the Neo-Granadian landlords, for whom the cost of a few stone columns was excessive, to pay, first, for a work of art carved in the shape of a baroque fountain, and then for a whole hydraulic system to bring water to the centre of the house.

The bath was an Hispano-Muslim tradition that was reluctantly brought to the New World by people who, according to the documentary evidence, didn't bathe very often. It is worth recalling the surprise expressed by the Castilian reconquerors of the Andalusian cities in the fifteenth century when they saw buildings whose function was as novel as it was strange to them: the public baths of Granada, Cordova, Jaén or Seville. Thus, the Spanish peasant saying:

"Water for horses, slops for pigs, wine and liquor for Christians (or, in the Neo-Granadian version, water for mules, slops for pigs and corn liquor for those on foot)."

It was in the tropical climates of the Cauca Valley and the temperate ones of the upper Cauca that the tradition of the open-air bath as an architectural element, and with it, that of elaborate systems of aqueducts, was revived in city and hacienda houses.

In the Cauca Valley, the sugar mill hacienda of *Piedechinche* (the current site of the Sugar Cane Museum) has a system of water supply for industrial use that dates back to the eighteenth century and is the best of its kind. *Japio*, another sugar mill hacienda in Cauca, includes a showy aqueduct on small masonry arches, and *Cañasgordas*, a cattle hacienda on the outskirts of Cali, has a covered bath incorporated into the main house, all of which confirm the importance and care that were given to water in rural buildings of western Nueva Granada. It is in the hacienda houses around Popayán where the handling of water as a construction material reaches an inspired level. The tradition of Muslim Granada, whose maximum expression is the architectural use, in the Generalife and the Alhambra, of fresh, cold water brought from mountain summits and guided by culverts and tanks in order to irrigate gardens, quench the thirst and wash "animals and christians" was continued, in a beautiful and utilitarian way, in the hacienda houses of *Antón Moreno, Calibio* and *Yambitará*. Sound and beautiful is that architecture which employs air and water to emphasize qualities of the natural space and give sense to constructed forms.

All of the above observations emphasize the futility or irrelevance of typifying or classifying Neo-Granadian hacienda houses on the basis of spatial, environmental, functional or simply climactic typologies. The author of this essay also yielded, in 1975,[34] to the attractive temptation, created by other historians, of relating the variety of climates in Nueva Granada to the architectural "types" of its hacienda houses, dividing them into "highland" and "hotland" ones. This is as entertaining (but as futile) as organizing them according to their location in places where it rains a lot or a little or where there are bats or insects. On the basis of the exhaustive inventories of rural colonial architecture that have been made during the past twenty five or more years, and the incontrovertible conclusion drawn from them, namely, that the same schemes of spatial ordering, the same technologies and the same volumetric results are to be found, without distinction, in the hacienda houses of *all* the regions and climates of Nueva Granada, with the exception of some small technical differences (or example, the use of different species of wood to mount the same structural systems), one has

Opposite page, *Fagua,* Cajicá, Cundinamarca. The complementary spaces defined by the surrounding walls establish the virtual space which frames the Neo-Granadian hacienda house.

to admit, in all frankness, that the theory of a causal relationship between house and climate lacks all sense or validity. This leads one, of course, to emphasize the tendency of the colonial landowner and builder to confront the climactic extremes of the country with identical architectural solutions, so that, in essence, they built the same house for the cold high moors of Boyacá or southern Colombia, the torrid tropical region of the Caribbean coast and the temperate-to-cool Andean high plains. The Spanish settler, accustomed to endure the fierce summers of Extremadura and Andalusia or the harsh winters of Castille and La Mancha, brought with him to the New World an architecture based on the elemental principle (or is it discipline?) that the deficiencies of a house can be compensated for by a tough stoicism. Thus it is that a good part of the reforms perpetrated on colonial hacienda houses during the late nineteenth and early twentieth centuries are tacit concessions to climate, that is, their intention was to raise the level of "comfort" (an idea that would have shocked the hacienda owner of the eighteenth century) by means of curious and usually unsuccessful attempts to either drive the heat out of the house, in the tropics, or keep it within, in the "highlands".

Another factor which further invalidates any kind of typological classification of Neo-Granadian hacienda houses has to do with formal considerations. These houses were created as a working tool and a refuge against the climate, not as works of art. They do possess, to be sure, an aesthetic dimension but this derives from and is subordinate to the first two features and therefore lacks an autonomous existence. The hacienda house does not have what might be called an "original form", much less a "final form". It begins in many ways (or in any way) and with many forms, including the temporary adoption of an indigenous hut, and during its active life is subject, by its very nature, to a gradual growth and decay that are obviously fortuitous or "untidy" from an academic point of view. It is destined, during its existence, to pass through the hands of successive owners who are going to destroy, restore, reconstruct and remake parts of the house in accordance with concrete necessities, temperamental whims and all sorts of ideas or accidents caused by nature or human folly. The house will grow in order to shelter more people and then be partly demolished when its dwindling number of inhabitants only need a room or two to live in amidst the terrible solitude of the countryside. Or destroyed when its destiny is to be abandoned and die.

A hacienda house whose history does not include constant changes and endless reforms would be a notable exception. There are innumerable hacienda houses that arose out of the firm intention, by their owners or builders, to enclose a central courtyard with four surrounding sections of building and were later reduced to a single pavilion, standing alone in the landscape, with a modest gallery on one face, because of the hazards of destiny, among them a sudden shortage of cash or credit. By contrast, there are others in which each generation of owners added "a bit more", charming and affectionate touches that create a warm or curious architectural "patchwork quilt".

It is as difficult to classify hacienda houses as it is to classify the faces of their inhabitants or builders. They were created in the exact likeness of their inhabitants and their idiosyncrasies, virtues and defects.

There has been a lot of debate, with regard to colonial architecture, about what is called "variety within uniformity ". Clumsy as this architect's term may seem, the fact is that the hacienda houses of Nueva Granada are always the same but always different. As are our own hands, which invariably have the same anatomy but an infinite variety of finger prints. We should not apply the old saying to them: "Seen one, seen them all". We would search in vain, through the whole of Nueva Granada, for two hacienda houses that are completely identical or situated in the exact same landscape. For a student of architecture this paradox of variety and uniformity is a matter of erudition, but a biologist would say that it is not only familiar but unimportant, given that what is vital and decisive in biology are not similarities but subtle differences, not typification but mutation.

The colonial hacienda house was not thought of in order to make concessions to the climate: very little thought was given to the question of achieving minimum standards of habitability.

Its rusticity or environmental harshness are an inevitable reflection of the implacable, though beautiful, nature which surrounds it. Luxuries of furnishing and decoration were bourgeois matters that were invariably reserved for the houses of the nearest city or the provincial capital. The extraordinary little palaces, that some Mexican hacienda houses are, belong to social groups that had much more economic power than the landlords of Nueva Granada and to levels of social conduct in which European ideas about architecture being a symbol of class and power reappear. And to people who had no problem in investing enormous fortunes in the most expensive of all luxuries – architecture. Nueva Granada would have, as corresponds to the landowning class that arose in it, generous and environmentally-likeable hacienda houses, but with no luxury other than their noble architectural lineage. Today, as in the past, to reside in a hacienda house in the Neo-Granadian countryside one must have the capacity to adapt to circumstances which are quite different to those which are now understood as habitability, an infinite patience and an affable and affectionate attitude: the architecture of the past requires it. We should not forget that the evocative charm and attractive atmosphere which country houses possess and never completely lose are to be found in the heart, mind and senses of the person who lives in or visits it and above all knows and understands it well, rather than in its architectural components. The essence of its possible grace as a constructed form does not reside in its unshakeable simplicity but in the complexity of the emotions or impressions which the perception of its spaces and their content may awaken. We should remember that the tough and humble implacability of the colonial hacienda house is not a drawback but its most notable virtue, being, as well, the architectural and environmental feature which allows it to enter and play an outstanding role in Latin American history. To the hacienda house, even more than its urban equivalent, we may apply the words of Antoine de St. Exupéry[35.]: *"…the palace of my father, where all steps had a meaning."*

Previous pages, opposite page, and right, the house of *Antón Moreno,* Popayán, Cauca. Among the haciendas near Popayán, the most inspired relationship between the house and landscape is, that of *Antón Moreno.* The singular mixture of harshness and gentleness in the architecture of the house is that which is shown by the surrounding landscape and the house arises from and depends on the latter. It reached its form during the last decade of the eighteenth century, after numerous earthquakes and alterations, although the hacienda had existed as a sugar-mill property at the end of the previous century and was later converted to cattle-rearing. *Antón Moreno* is a synthesis of basic schemes of spatial ordering commonly found in the architecture of Neo-Granadian hacienda houses. It elegantly combines a main patio, crossed by a branch, on small arches, of the aqueduct complex of the house, with exterior galleries open to the surrounding panorama. It includes a chapel annex joined to the house by the section which lodges the outbuildings. The chapel has an unusual altarpiece painted on canvas in false relief.

Previous pages, left and opposite, *Calibío,* Popayán, Cauca. It would not be surprising to learn that the same hands had built the houses of *Antón Moreno* and *Calibío,* which lie at opposite ends of the outskirts of Popayán and were built between the seventeenth and eighteenth centuries, though mostly in the latter. *Calibío* is an "up and down" house characteristic of a plain where one must ascend in order to look out at the landscape, while *Antón Moreno* is a hill-top one that is already situated in the heights. Benefitting from a larger budget, *Calibío* assumed larger and more elegant proportions than those of its fellows in Popayán, and has the tone of a great house in the city, super-imposing an upper storey upon a ground floor which perhaps is more architecturally refined than that of *Antón Moreno* but less intimate and evocative. *Calibío* displays the greatest luxury of construction and most refined technological level seen in the Cauca hacienda houses.

The celebrity of *Calibío* as the setting for important political and military events during the nineteenth century has, to a certain extent, dulled its architectural significance and the outstanding technological quality of its construction. *Calibío* is an archetypal example of a genre of building that gradually left behind its roots in Andalusia or La Mancha to assume a new identidy in the land of Nueva Granada. While it is true that tiles are of an Arabic kind, as are its roof frames, they are placed upon a base of interwoven canes in the indigenous manner. Its timbers, adobe blocks, bricks, lime, clay and canes are from the region and have been worked and joined together according to regional techniques. In a certain and real way, the house arose from the place where it was built, even though its character and spatial organization date back to ideas thought up,centuries before, in the remote lands of other continents.

Left and above, *Calibío,* Popayán, Cauca.
The great galleries of the upper storey of *Calibío* are practically identical to those which may be found, even today, in country areas and cities of Spain, from La Mancha to Cordova, but they were created as they are and in no other manner because the landscape of Cauca imposed itself upon the memory and intuition of its builders. For this reason they are, like the rest of the house, inseparable from the place where it stands. The hacienda house is a setting for a mode of everyday existence that cannot (and never was intended to) be the same on both sides of the ocean. In *Casa Colonial* we read that: *"The Neo-Granadian houses will show a ...mixture of traits that are half-Castilian and half-Andalusian, that is, something which was never seen either in Castile or Andalusia. Thus was created the original nuance ...of the flourishing, on the new continent, of an architectural genre that is as old as humanity itself. Its lineage was always clear in this, but its intimate relation with the land where it was built is also evident".*

Previous pages, left, and right page, *Piedechinche,* El Cerrito, Cauca Valley. (The present Sugar Cane Museum). With the introduction, well into the colonial period, of the cultivation of Asian sugar cane, the haciendas of the Cauca Valley became divided into two genres. The haciendas which produced sugar soon became more profitable than those dedicated to cattle. Some of the more prominent sugar-mill haciendas of the region, like *El Alisal,* have been destroyed recently. *Piedechinche* has been turned into a museum which demonstrates the region's agricultural history and has the merit of revealing the spatial organization of an industry that revolved around a hacienda house –a better fate than demolition or modernization. Note how the landlord's house of *Piedechinche* shows spatial similarities to Cauca haciendas like *Coconuco, Pisojé* and *Yambitará*. These include the elevation of one section in the manner of a lookout balcony and the use of very wide exterior galleries in the upper and lower floors, without covering the whole of the façades.

Opposite page, and right, *Piedechinche,* El Cerrito, Cauca Valley. In most of the "old" sugar mills of the Cauca Valley one sees versions of the industrial technology of the nineteenth or twentieth centuries. In *Piedechinche* the mill retains its colonial character, with some reforms from the nineteenth century. The ovens and the chimney date from the same epoch, but the sugar cane shed and storage sheds are older, as is the aqueduct, which is partially raised up on brick arches. Contrary to what happened in the olive haciendas of Andalusia, where the manor house was integrated with the adjacent industry, the sugar-mills of the Cauca Valley give the impression of having been located, by accident, in the vecinity of what are, strictly speaking, the hacienda houses. The industrial architecture of the colonial period made use of structures that were devoid of any concession to the possible gracefulness of constructed forms, thus creating a marked environmental contrast with the nearby hacienda houses. A mill could be situated in any place where there was water, but a hacienda house needs a special setting.

Left above, below, and opposite page, *El Puesto,* La Ceja, Antioquia. Built at the middle of the eighteenth century by the royal standard bearer Felipe Villegas y Córdoba, the house displays volumetric and technological features that were also present in houses from the same period in Cauca and the Cauca Valley, thus invalidating typologies which associate distinctive architectural forms with the different regions of Nueva Granada. Note the volumetric heightening in the second floor, in the manner of a projecting (and not a recessed) balcony and the exterior balcony on the ground floor. It is rare to find hacienda houses in the region of Antioquia which conserve, as this does, their setting and surrounding space and have not been altered to the point where they have completely lost their original architectural components. The presence, in its landscape, of the house of *El Puesto* is exemplary.

Left above, below and opposite page, *Saldaña,* isolated chapel, Saldaña, Tolima.

The reason why annexed, free-standing or isolated chapels were built near rural houses in Nueva Granada had to do with the need to provide religious services to parishioners who lived in neighbouring farms, or hamlets which lacked a church: this, apart from attending to the needs of the hacienda itself. Around the chapel a space was needed for open-air processions and other ceremonies. The house of the hacienda *Saldaña* went downhill in the nineteenth century and has lost, in the twentieth, a large part of the character given to it in the republican period, but the colonial chapel still remains, which was apparently built at the end of the eighteenth century. Its naïve baroque features, which were obviously done "at sight" by an anonymous builder, are limited to two half ovals that have been cut into the diagonals of the pediment of the façade, thus producing a silhouette that mixes straight and curved lines, and, in the interior, a singular faceted altarpiece in masonry. The chapel is what might be called "recondite rural Neo-Granadian baroque".

Previous pages, opposite page, right above and below, *Japio,* Caloto, Cauca. The house and mill (for chocolate beans) of *Japio* reached its maximum point of original development in the second half of the eighteenth century. The colonial house, with a compact ground plan and volumetry similar to that of *La Merced* and *El Hato,* is surrounded today by formal gardens which date from the end of the nineteenth century, as is the extensive aqueduct upon brick arches, a work of hydraulic engineering from the republican epoch. The house has had an unlucky history in the twentieth century. Its free-standing chapel, which had been used as a storehouse, was demolished in 1917. The custom of knocking down rural chapels became fashionable in the region, where those of *Cañasgordas, Pisojé* and *Puracé* were also destroyed. The house was modernized in the nineteen-seventies, depriving it of the parts that made up the corners of the building, which unfavourably altered its outward appearance, and furnishing it with an unusual new stairway hat "combines with" the original one.

Left and opposite page, *Cañasgordas,* Cali, Cauca Valley. The *chorro* or bath for the landlords has a keyhole shape, which is to be seen today in archaeological excavations done in the groups of artisanal houses of the Islamic epoch discovered in the Alhambra of Granada, which date from the eleventh and twelfth centuries. An architectural memento that was renewed six hundred years after the period of its origin.

The house of *Cañasgordas* dates from the beginning of the eighteenth century, with numerous reforms and additions in that and the following century. It conserves its mill but its free-standing chapel was demolished, as often happened at the beginning of the twentieth century. As with most sugar-mill hacienda houses in the Cauca Valley, it has a compact ground plan (with extensions), provided with the usual large upper and lower floor galleries. *Cañasgordas* appears to have been started in the seventeenth century as a low house with a single section, which then grew with the passage of time. The dominant spaces in *Cañasgordas* are not its precints but its galleries, where a large part of the daily life of the house took place.

Left and above, *El Abra,* Zipaquirá, Cundinamarca. It may be defined as an example that lies on the border between a modernized colonial hacienda house and a present-day country house furnished with a traditional volumetry and setting. *El Abra* belongs to the ambiguous genre that might also include the houses of *Fute, Canoas, Cortés, Buenavista, La Conchita* and other country residences in Cundinamarca. The house conserves the more or less original volumetric relation with the splendid place where it stands. The chapel and the atrial stone wall are modern intrusions (from the nineteen-sixties), the latter having notably affected the organization of the spaces which surround the front of the house.

Left and right, *El Abra,* Zipaquirá, Cundinamarca. What stands out is the contrast between the elegance of the atrium with stone kerbs and pinnacles (seen in the plazas of Baeza or Ubeda in Andalusia) and the discreet but evocative modesty of the secondary patios which retain their traditional character.

Above and right, *Tilatá,* Chocontá, Cundinamarca. Apparently dating from the eighteenth century, *Tilatá* displays an attractive volumetry which indicates that it grew by stages. Although there is no lack of the utilitarian structures which inevitably arise alongside almost all of the hacienda houses of Cundinamarca, the house conserves, to a large extent, its remarkable positioning within a terrain which has (as of 1997) not yet been invaded by quarries, country clubs or factories. Even more important, the spatial cover of walls and adobe enclosures, that are a vital complement to the house, still exists.

"…the hacienda houses… are always the same but always different. As are our own hands, which invariably have the same anatomy but an infinite variety of finger prints. We should not apply the old saying to them: Seen one, seen them all…
G. Téllez.

Opposite page, right above and below, *Pisojé,* outskirts of Popayán, Cauca.
It appears to have been constructed (replacing a humble hut on the same site), in the second half of the eighteenth, by those who also built the hacienda houses of *Coconuco* and *Yambitará* in the same region and *Piedechinche,* in the Cauca Valley. This group of houses shows an identical constructive technology and structural systems, as well as a spatial organization around a compact ground plan with generous "unfinished" galleries on its longitudinal fronts. They also share a partial elevation in the form of a lookout balcony.

Pisojé is situated in a spot which brings together practically all of the elements which compose the region's geography: gentle hills, steeper mountains, an adjacent river, a distant panorama of a high range with volcanoes and snow-covered peaks and dramatic sunsets. Repeated observations of the spectacle of the precise architecture of the Cauca hacienda houses might lead one to conclude that the poetical sense of place which distinguishes them turned, in a way, into a conventional feature. The reiterated and invariably sure touch shown by hacienda owners and builders, when they sought to take maximum advantage of the environmental possibilities of their sites, became a custom and tradition with the passing of time.
Pisojé was not exempt from the waves of vandalism which periodically wash over the Colombian countryside. While the house was saved from a drastic modernization, its free-standing chapel, like the ones in *Antón Moreno* and *Calibío,* was knocked down in order to install a "bath" for cattle, with the argument that the church of *La Jimena* only lay less than half a league away and it was just as easy to save your soul in one place as the other.

Left and above, *Los Aposentos,* Chocontá, Cundinamarca. Various hacienda houses in Cundinamarca bear this name, which is also frequent in Castile and Andalusia. At the beginning of the eighteenth century haciendas called *Aposentos* existed in Simijaca (illustrated in previous pages), Sopó and Chocontá (which is shown here) and several others appeared in the region during the nineteenth century. As in the case of *Casablanca,* the reason for the proliferation of certain names for rural properties is much the same as for the popularity of names like Pedro or Juan among the Spanish settlers of Nueva Granada. *Los Aposentos,* on the outskirts of Chocontá, appears to have been originally built in the first half of the eighteenth century, although the hacienda itself was formed in the middle of the seventeenth century. It lacks the academic excellence of the house of the same name in Simijaca, belonging, instead, to the genre of "classic" savannah houses. As such, it displays splendid modular proportions in the colonnades of the galleries shown here and a delicate "understanding" with the surrounding landscape. It appears that the original house only consisted of the longitudinal section seen on the right, which would make it similar to the type of house with a compact floor plan with large galleries on its longitudinal sides. This confirms the idea that a spatial organization for hacienda houses that is specific to (or typical of) a particular climactic zone or region does not exist, given that houses with a compact floor plan are to be found from the south of Nueva Granada to the high plains area of Cundinamarca and Boyacá. Fortunately, the decades-long growth of the house did not jeopardize the volumetric limits and formal continuity which make *Los Aposentos* one of the purest examples of the Cundinamarca hacienda house.

"Hacienda houses… were created as working tools and refuges against the elements, but not as works of art".
G. Telléz.

Left and opposite page, *Altamira,* Tenjo, Cundinamarca.
A good example of the environmental dimension and additional meanings which a hacienda house acquires with the passing of time, when its utilitarian character gives way to other reasons for its conservation. *Altamira* appears to be one of the many names given, in the nineteenth century, to properties and houses which arose out of the dismembering of the oldest haciendas in the savannah of Bogotá. In a house with such modest dimensions it is apparent that the physical size of the buildings had little conceptual importance for the builders of the colonial period. Their invariably sure formal and environmental touch could be equally applied to a minimal or an enormous format, without their needing to vary the materials employed or the constructive techniques. In the minimalist architecture of *Altamira* the tiled roofs cannot be more expressive (or lower), nor the patio more reduced or more evocative.

Left and above, *Casablanca,* Sopó, Cundinamarca. In common with other hacienda houses of the colonial epoch in this region, it bears the unmistakable features of the predominant dimensions of its tiled roofs and the modest height of the perimetral galleries in the façades. As in *Chaleche,* Sesquilé, *Los Aposentos,* Chocontá or *Fagua,* Cajicá, the roofs and tiles of *Casablanca* establish the presence of the house in the landscape and completely dominate its volumetry, given that it is only in the middle of the nineteenth century, that the façades (galleries and walls) recover their importance. The builders of Santa Fe and the savannah of Bogotá were talented, above all, in constructing large and beautiful roofs which could, when necessary, shelter a house. Note the generic resource of the section in two stories, enhanced along the lines of *Fagua* and the hacienda houses of Cauca.

Above and right, *El Hato de Córdova,* Facatativá, Cundinamarca. A good example of the numerous farm or hacienda houses built in the west and south of the savannah of Bogotá between the end of the Colony and the first decades of the republic and maintaining the traditions of spatial organization and architectural language characteristic of the earlier period. The houses of this epoch have an ambiguous appearance. In general they display a carpentry of doors and windows characteristic of the second half of the nineteenth or the beginning of the twentieth centuries, applied to walls that might be attributed to any epoch. The relation between house and place is, in these examples of a transitional character, more generic and undefined than that which is seen in the houses of the end of the seventeenth century or the beginning of the eighteenth.

Following pages, *El Chacal,* Tenjo, Cundinamarca. It resulted from the subdivision, at the end of the seventeenth century, of the enormous property of *Tibabuyes,* in the west of the savannah of Bogotá and the surroundings of the town of Tenjo, Cundinamarca. The house pictured here might have been constructed immediately after the formation of the hacienda, that is, at the beginning of the eighteenth century. *El Chacal,* in turn, was subdivided in the second half of the nineteenth century, which leads us to suppose that the decorative republicanization of the colonial house took place in the last decades of that century. The house of *El Chacal* retains its beautiful siting, dominated by the silhouette of its tiled roofs, but now devoid of most of the surrounding enclosure and adobe walls, that is, of the limits of the complementary spaces within which its architecture is set.

Above and right, and opposite page, *Polmerán,* Sotaquirá, Boyacá. This is another example of ambiguous architecture. It might be equally taken for an entirely modern house, built with some traditional constructive elements, like the roof frames in the "par y nudillo" technique (see note No.30), or a relatively old house which has been extensively enlarged and reformed. The spaciousness of the interior patio and the manufacture of the roof frames illustrated here, which are different from the artisanal usages of the region (examples of which are shown in other parts of this book) confirm this. Be that as it may, in this architectural genre and given the present circumstances, the end (to achieve a pleasant and beautiful atmosphere) justifies the means.

Left and opposite page, *La Loma,* outskirts of Santa Fe de Antioquia. In these and the following pages we see two houses from the same region and with the same name. Is it the house which is encrusted in the landscape or the landscape which invades the house? The original sections of the house of *La Loma* ("the hill"), a name which is very frequent in such an extremely mountainous region, date from the beginning of the eighteenth century. The exquisite integration of the roof tiles, corridors and surrounding walls with the natural space around them is the result of a profound understanding of the essence of the landscape and of a prolonged and affectionate concern for the constructed forms and the vegetation. These photos capture the relation between, not only the house and the place, but also – and perhaps most important, – man and the countryside.

Opposite page above and below, and right, *La Loma,* the second house with this name, outskirts of Titiribí, Antioquia.
The placing of the house in a setting with an abrupt topography borders on the theatrical. The house invades the place, making use of the same architectural resources that might have allowed it to stand in the flat landscape of the Cauca Valley or in the high plains of Boyaca, among them, ample tiled roofs and exterior galleries, as well as the traditional interior patio. The house on a hill on the outskirts of Titiribí was, like the first one, built in the eighteenth century, in that region of Nueva Granada where one finds most of the dual-purpose haciendas that are analogous to those of Mexico, that is, mining and agricultural at the same time. Many of them, because of the great difficulty of access from the nearest towns, acquired pseudo-urban characters, furnishings and usages, given that their owners preferred to live in them for long spells, rather than undertake repeated and hazardous journeys to the province's cities. For this reason, the interior shows the eclectic treatment of the republican period.

Left and above, *Rastrojogrande,* La Virginia, Antioquia. Constructed, at least in part, during the eighteenth century, the house underwent the nineteenth century transformations usually found in the rural architecture of Antioquia, as well as quite a few amplifications of wall openings and other more or less inevitable modernizations. The longitudinal front of the house conserves a pleasing rusticity, which, while it is not original, is at least fairly old. Note the long eaves and, in the foreground, a mounting block for horses or carriages. These are still occasionally seen in some Andalusian farms which breed or keep horses.

The Colombian hacienda house

In Colombian territory, the end of the colonial period may be situated in the years between 1819 and 1830. Colonial hacienda houses built up to that time may be correctly referred to as *Neo-Granadian,* that is, built in the territory of the *Nuevo Reino de Granada* (the New Kingdom of Granada). From these dates onwards, both new buildings and transformations made on ones from the previous period may accurately be called *Colombian,* given that, chronologically, the formative period of the new nation begins there and with it the great transformation of all that was built in the countryside of what would be successively called the *Estado de Nueva Granada* (the State of Nueva Granada), *Confederación Granadina* (the Granadian Confederation), *Estados Unidos de Colombia* (the United States of Colombia), and finally, in 1886, the Republic of Colombia. A description of the complex historical and political process which marked the transition, from a colony to an independent nation, of what had been the Spanish "overseas province" is beyond the scope of this essay, but we can explain something of what happened in rural architecture during this time.

The socio-economic changes resulting from this political transformation had, at most, a minor impact on the reality of the countryside. Throughout history, the rural world has traditionally been unwilling to accept social change. The thesis that the only concrete results of the wars of independence of the nineteenth century in Nueva Granada were the replacement of one set of bureaucrats by another and the passing of political power from the Spanish ruling class to its Neo-Granadian equivalent has become ever more convincing in the light of modern researches that cast doubt upon the old-fashioned and overly patriotic "official" version of Colombian history. The stratification of social classes, disguised under a different label, largely remained intact. The regime of land owning and the relative standing of the peasant-farmer, the share-cropper, the colonist and the landlord did not significantly change. The economic systems of production and supply continued to follow the guidelines laid down in the colonial period, with only a few slow and superficial changes. The wars of independence led to the disappearance or exile of many Spanish and creole members of the landowning and middle classes and they began to be replaced by new social groups which did not noticeably modify traditional attitudes about the ownership and use of land.

The dividing-up of large rural properties continued, as in the late colonial period, but it has never been demonstrated that this process notably accelerated during the republic. One proof of it is that even today (1997) the large estate (*latifundio*) and the narrowest small-holding (*minifundio*) continue to coexist in the Colombian countryside. In regions like the Caribbean coast, Cauca, the Eastern Plains, Huila and Tolima, most of the *latifundios* of colonial origin survived throughout the whole of the nineteenth century, despite the fact that many had been split up when they passed from the hands of Spanish to those of creole landowners. The socio-political instability in Colombian territory in the nineteenth century, marked by many civil wars and inter-regional struggles, slowed down or stopped the economic development of areas which had had a limited agricultural production during the colonial period. A brief but bloody civil war was all that was needed to decimate the rural workforce of a whole region and the depopulation of the Colombian countryside thus became an inevitable and uninterrupted phenomenon, together with the resulting concentration of population in cities and towns. In short, the authoritarian colonial system (even with all its faults and limitations) imposed a relative economic order that was not replaced, in Colombia, by a protocapitalist or other one. The overall impression one gathers from an examination of the many studies that have been made of the period is of what appears to be a general disorder, in the midst of which one perceives the continuance of certain colonial tendencies, such as the monopolization of unproductive lands, a non-technified agriculture, the parasitic role of middlemen in the marketing of products, the isolation and lack of transport in cattle-rearing regions, and so forth.

The appearance in Colombia of what was known, in Europe and the United States, as the "industrial revolution" was late, slow and confused. Among the new

Opposite page, Patio of the house of *Gotua,* in Iza, Boyacá. Built during the last years of the eighteenth or early years of the nineteenth centuries.

phenomena that decisively affected the infrastructure of the Colombian countryside there was the development, always difficult or truncated, of the railways. Railway lines, together with large-scale open sky mining, changed the face of the landscape of many Colombian regions forever and not necessarily for the good: more than one colonial hacienda house found its surrounding landscape turned into a mini-desert as a result of it. The capitalist dream was that the railroad would open up regions which were inaccessible before and facilitate the transport of an avalanche of products that would soon flow out of the immense Colombian cornucopia. That the real history of this energetic effort to heroically penetrate the nightmarish landscape of the country with a rail network turned out, in fact, to be very different from the theories proposed by political leaders is another matter. The routes of more than one Colombian railway were only laid down in order to satisfy the big landowners, who (at times being the political chiefs of the republican epoch) made sure that the trains passed by their colonial haciendas and that they would be paid exorbitant prices for the big strip of land needed for the line and for rights of way.

The following apocryphal story about the route chosen for the Girardot-Tolima-Huila railroad is characteristic. It has to do with the building of a line through the estate of a distant Colombian descendant of some unknown Spanish colonial landlord whose name only God remembers. The routing of the line was in the hands of the eldest son of the estate owner, who, in accordance with the new practice of the ruling class of the young republic, had been sent to the United States to study engineering (or whatever that business of working with theodolites and paper drawings was called). "What an idiot this boy of mine is", the estate owner complained to his friends. "He's decided to send the railway line through the middle of the living room of the old hacienda house." A while later, the young engineer, after redesigning the route, announced his conclusions to the head of the family: "Father! The line will no longer run through the living room of the old house! I managed to put it through the peons' quarters..."

In the inevitable triple clash between the new technologies, "progress" and tradition, it was obvious which would come out worst. More often than not, the old hacienda houses lost the argument, first with the railways, then, decades later, with the roads and lastly, they were asphyxiated by the apocalyptic urban growth of Colombian cities.

The republic inherited what the colony had left it, geographically and economically speaking, including the fundamental fact that the lands which were most suitable, in terms of climate and population, for agriculture and cattle-rearing had a greater concentration of well-established haciendas and therefore, of hacienda houses that were architecturally distinguished. Such is the case of the high plain of Cundinamarca and Boyacá, Cauca and Cauca Valley, the regions responsible for 80% of the colonial hacienda houses illustrated in this book. Nevertheless, the rule has some exceptions. Regions as apt for cattle-rearing and/or agriculture as the Valle de Upar, the prairies of the Magdalena, the so-called "Momposina depression", and the prairies of Casanare, Tolima and Huila also showed a certain level of exploitation and development but they do not conserve hacienda houses that are as notable as those which survive in the Cauca Valley or Boyacá. The reason is simple: colonial landlords did not always think it convenient or necessary to establish a permanent residence in rural areas with an extreme climate when it was just as easy and cheap to live in a more or less nearby city or town, for example, Mompox. The *latifundios* in the region around this city only had a palm-thatched, mud-walled hut for the ranch foreman and cowhands, as Orlando Fals Borda points out in his *Historia Doble de La Costa*[36]. As we have already shown, even in the regions of the country where the best colonial hacienda houses are found, their appearance was late and slow.

The case of the region of Antioquia is more ambiguous. Having a mountainous geography which favoured mining much more than agriculture or cattle-rearing, its colonial economic development was different from that of adjoining areas, like the prairies of the Caribbean coast. There was a markedly divided tendency among its settlers, perhaps because its rugged topography made communications and land transport difficult, as well as the fact that many of its inhabitants were devoted to widespread commercial activities. Thus, the tendency was to reside in towns or cities like Santa Fe or Rionegro but also have hacienda houses, although these never came to boast of such a significant architecture as those, for example, of the Cauca Valley. Because of the need to make long and hazardous journeys from a city to a hacienda lost in some corner of the Antioquian mountains or vallies, it was convenient for them to maintain, at both ends of the route, a place that allowed them to rest and carry out their day-to-day activities.

The advent of the republic little changed the situation described above so far as Colombian economic geography was concerned. Even nowadays the regional distribution of agricultural and cattle-rearing haciendas continues to follow the patterns laid down in colonial times, despite the diversification of crops and the economic development characteristic of the twentieth century. There is, however, one notable exception to the above rule: what has come to be known as the "Antioquian colonization" of the central (mountainous) part of the country, starting in the second half of the nineteenth century, that is, at the height of the formative period of the Colombian nation and in the face of every imaginable kind of political and socio-economic problem. The "Antioquian" case is an understandable example of regionalism, given that a majority of the colonizers who decided to "open" rugged and unpopulated territories were people from the central and southern region of Antioquia. The architectural expressions of what in reality came to be one colonization imposed upon a much earlier but far more tenuous one were largely urban. The rural parts of the centre of the country saw new agricultural activities arise, apart from the already traditional ones. Thus, although the cultivation of coffee was already known at the end of the colonial period, it gathered an unforeseen and extraordinary momentum in the nineteenth century, within the geogrphical boundaries of the regions of Antioquia and what would eventually come to be known in the republic as the "old" department of Caldas. Those regions possessed ideal climactic and geographical conditions for the selective growing of coffee, which was of African origin and spread through the New World by way of Brazil. In only a few decades, coffee became the dominant socio-economic factor in the centre of Colombia, helped by the emergence of international markets for the product and the creation of large accumulations of capital derived from its growing and marketing.

It is worth pointing out here that the inclusion of only two examples of the coffee hacienda or farm house in this book is due to the fact that, because of its considerable merit and interest, the subject would require a separate study, which has, in fact, been made in *La Arquitectura de la Colonización Antioqueña*[37], a series of books edited by the architect Néstor Tobón. That aside, the origins and development of the coffee farm house were quite different from those which played a formative role in the farming culture of the colony: in addition, they occurred in regions that were barely settled and exploited during Spanish rule. The cultivation and marketing of coffee in Colombia is a socio-economic event of a proto-capitalist nature that took place in historical circumstances that were completely different from those which applied at the beginning of the colony. The colonial hacienda derives from medieval notions about the ownership and exploitation of land. The coffee farm lacks such an origin. We might say that the rural architecture associated with the colonization of the centre of the country in the nineteenth century is a singular and atypical eddy within the general historical current which leads from the colony to the republic. And add that the house of a coffee farm in Antioquia or Quindío cannot be considered simply as a branch of the family tree of colonial hacienda houses, but, rather as an architectural species which is original to its region, although having certain inevitable and highly generic spatial features in common with the hacienda house and shared with other regions of Central America and the Caribbean which also grew coffee. The landlord of a hacienda of the seventeenth or eighteenth cen-

tury in Cauca or Boyacá has very little or almost nothing in common with the coffee-grower of the nineteenth or twentieth century in Quindío. The place which they occupy or role that they respectively play in the socio-economic systems of the colony and the republic could not be more different. Therefore, their rural houses reflect these considerable differences.

During the period which has been dubbed the "republican" one, that is, running, in Colombia, from the end of the first third of the nineteenth century to the third decade of the twentieth, many - though not all - hacienda houses of the colonial period went through, to one extent or another, a process of formal and technological transformation that could never have taken place during the colony. Beyond this, it is obvious that a certain number of *new* hacienda houses would appear, that is, ones that were built after the first third of the nineteenth century in a formal and technical language that was decisively different from what had been established by colonial traditions. These two architectural phenomena make up what might be called the distinguishing mark of the epoch.

This transformation, above all of interiors, but which eventually affected the external volumetry of hacienda houses as well, began in the nineteenth century and its origins were very different from those which had applied to building in the colonial epoch. The eclecticism - the selection and mixture of architectural elements or motifs from numerous and variegated sources - which replaced the formal unanimity that was dominant before then did not have a technological or traditional origin but a commercial and superficially aesthetic one. The eclectic hotchpotches which reached Colombia were not the product of a social consensus or deep-rooted customs but of economic impositions by the industry and "free trade" of some European countries (France, England, Italy and to a lesser extent, Germany and Spain itself), as well as the United States. Rural architecture now became a channel of consumption for the mass-produced articles of the "industrial revolution".

Some commentators have tried to see, in the simple commercial rush to impose the use of certain decorative and construction materials on nineteenth century Colombia, a supposed search for a new social and political identity related to a vague architectural symbolism. They have surmised aesthetic tendencies and currents in the merely accidental fluctuation of the new and frivolous phenomenon of fashion, transposed willy-nilly from dress to architecture. All of this amounts to little more than a good-natured historiographical and critical attempt to "improve" (or take too seriously) architectures which are only a little better than mediocre and superficial, or see profundity in what, even with the best will in the world, does not go beyond the simply pleasant or amusing. The colonial hacienda house, like its town or city equivalent, was modified with the introduction of flat ceilings made up of heavy lumps of lime and sand smeared on a base of cane or guadua bamboo, and in the resulting surfaces there arose mouldings, rosettes, medallions, cornices and anything else that might occur, indiscriminately, to those who sold and applied decorations in plaster ("the poor man's marble") or papier-maché. The "poor" rafters which made up the framework of the colonial roof were thus hidden behind a bourgeois mask and the humble and beautiful interior spatiality of the hacienda houses was lost. The rustic surfaces of interior walls were primed with "carpenter's glue" and then the universal finish of wall paper ("the decorative mural of the petit bourgeois") was pasted on with a starch glue. "Persian" carpets replaced the esparto-grass mats laid down on the adobe or brick floors. Elegant *parquet* floors in French woods were placed on top of (or replaced) the modest brick or rustic plank ones. Then came the invasion of the exterior facades by rainwater channels, metal cornices, downpipes, etc. in tin, "the poor man's bronze".

In short, rural architecture could not escape the "bourgeois-fication" resulting from industrialization nor the underdeveloped world's liking for the left-overs (or "fag ends") of aesthetic modes that were popular abroad. It was a matter of a complicated transition from artisanal building to a mail-order architecture, which could be directly bought, through the right store, from Europe or the United States. Nor should we omit the astonish-

Opposite page, *El Salitre,* Bojacá, Cundinamarca. Extensively reconstructed and reformed at the end of the nineteenth century.

ing countryside transformation of the latrine in the "backyard" into a French "bath room". Or the consequences of the appearance of such exotic technologies as galvanized iron or earthernware pipes. Or the effect, of the importation of machines to cut wooden mouldings and bend iron rods, on the hacienda house's modest doors of poorly planed planks and windows with cowhide instead of glass.

The entrance of the colonial country house into the modern epoch was like the rude and clumsy invasion, by a backwoods landowner dressed in a poncho and straw hat, of a genteel literary evening where all of the guests were wearing French frock coats and English bowlers. By the end of the nineteenth century the old hacienda house, after having been a noble, efficient and humble working tool, had become a symbol or emblem of social class. We must not forget, of course, as was shown in a previous chapter of this book, that the history of hacienda houses consists of the almost continual and indefinite formal and dimensional changes that were made to them, through additions and removals, and therefore we can not assign "final forms" to them or speak about supposedly "original" buildings. So that it is only logical now to assume that the transformations that took place in them in the republican epoch are simply episodes in their successive metamorphosis. Critical judgements or questions of taste relating to the aesthetic or atmospheric qualities of different hacienda houses of the colonial or republican epochs are quite a different matter.

The changes described above were by no means exclusive to Colombia. The same process, but at a much higher qualitative (and economic) level, was also seen in Mexico and Brazil, stimulated by the formation of social groups whose aristocracy and power was based on an enormous commercial and industrial wealth whose scale was international. We have only to look at some hacienda houses of the republican period in Colombia (or colonial ones that were republicanized to the last detail) to see, for example, in *La Julia* (Cauca Valley) or *La Ramada* and *El Corzo* (Cundinamarca), which are republican remodellings of colonial houses, the magnitude of the transformations made during the end of the nineteenth century and the beginning of the twentieth. Certain examples of wholly republican hacienda houses, such as *La Industria* or *Piedragrande* (Cauca Valley) and *Buenavista* or *El Noviciado,* Cundinamarca[39], all of which replaced colonial buildings or were constructed alongside them, allow us to establish the existence, in them, of shared architectural features. The most notable is, without doubt, the abandonment of the traditional scheme of spatial organiation, with its geometrical informality created by familiar customs, concrete necessities and everyday habits, in favour of volumetric arrangements and an organization of living spaces based on the imposition of abstract geometrical principles like a symmetry with respect to a single central axis, in two or three dimensions. It is rare or exceptional to detect any sign or intention of symmetry in a colonial hacienda house.

What sense or utility would a landlord of the seventeenth or eighteenth century have found in a house which had the same number, size and arrangment of rooms on both sides of an *imaginary* line that went through the middle of the courtyard and the stables? Nevertheless many of them and almost all of the republican epoch nowadays boast of approach roads which, instead of winding around the house to enter "by the back", as happened traditionally, run directly to the centre of the house's main front and require a stairway and an entrance hall in the front corridor, which decisively break up its physical and atmospheric continuity. It is worth looking, in this respect, at the main entrance of the house of La Sierra (*El Paraiso*) and the refined symmetry of the façade of *La Industria*. There is no doubt that the notion of "elegance", of formal refinement, entered, in a vague and indefinable way, into the history of Colombian hacienda houses.

In a colonial house like *Fusca,* which was not interfered with spatially, during the republican period, it is still possible to follow the original approach on horseback and reach the house by the back corridor. You then come out into the central courtyard, skirt round it and pass through the main hall to come out, at last, to the front gallery, from which one sees the panorama of the savannah of Bogotá. In *El Paraíso,* on the other hand, the approach route turns its back on the magnificent view of the Cauca Valley, without passing through and perceiving the house beforehand, in a way that is suitable for a tourist but not someone who actually resides there.

In hacienda houses dating from the second half of the nineteenth century, like *Buenavista* (Cota, Cundinamarca), which was built at a certain distance from the colonial house of the same name (which is now unrecognizable or practically non-existent), the spatial scheme was established deliberately and from the start on the basis of a longitudinal axis and a rigorously symmetrical facade, as were the spaces of the original reception rooms before they were unfortunately modernized to create a single living space. In this case, symmetry is not something imposed on what already exists but, rather, an original feature, created by a bourgeois of a high social class whose cultural level was somewhat cosmopolitan and who had a "snobbish" and novel idea about the new house to be built in *Buenavista.* Instead of a hacienda house, or as well as one, this had to be a *country house,* with the elegance, surprising in the savannah of Bogotá, of an avenue of cypresses brought from France, a porch with a wrought-iron grille and a courtyard for carriages, also symmetrical, in front of the main entrance. The aesthetic and social transformation of the hacienda house thus reached its high point in the last three decades of the nineteenth century.

Adapting a hacienda house, whether republicanized or not, to the modern conditions of the Colombian countryside brings with it an insoluble architectural problem. It was possible for a colonial house to advance into the republican era because the superimposition of implicit architectural elements and criteria could take place without completely erasing or violently contrasting with what already existed or had come long before. By contrast, the modernizations done to houses in the savannah of Bogotá, Boyacá or the Cauca Valle in the past thirty years have led to all sorts of contradictions, such as the absurd attempt to reconcile the preservation of traditional living spaces half-lit by small windows and the destruction of thick adobe walls in order to introduce the enormous windows of modern architecture. This brutal alteration of the relations between the interior and exterior of these old houses is, allowing for a bit of exaggeration, the equivalent of a marriage between an ageing Hollywood actress and a young gigolo whose only virtue is his good looks. The aesthetics of these unfortunate encounters between the past and present are extremely depressing.

Without doubt, the worst thing that can happen to a hacienda house during these current modernizations is the destruction or defiguration of its setting or immediate surroundings. This affects the most important characteristic of the hacienda house, which is the relation between the house and the natural space which surrounds it. The cutting down of ancient trees and the demolition of surrounding walls are the main examples of vandalism done to the surroundings of the hacienda houses. The first takes place on the pretext of "clearing" an area around the house in order to install a "designed" garden and spaces to park cars. The second, to eliminate the obstacles which "prevent one from seeing the view". This is the complement of modern goldfish bowl windows, given that the opening of large gaps in the façade often reveals the close and annoying presence of ancient enclosure walls, which certainly prevent one from appreciating the sight of the Freisian cows grazing some hundreds of metres away. It is true that a very few hacienda houses, a small minority, have managed to miraculously conserve their surroundings, and others now have beautiful modern gardens around them, but an enormous number of them have lost, forever, the features and atmosphere of the countryside they live in. Without this, the hacienda house is just another building, without much character. Perhaps the "best" and most painful examples of the above are the complete destruction of the surroundings of the house of *Santillana* (Tibabosa, Boyacá) by a highway and sports complex and the mutilation of the beautiful enclosure walls of the house of *Suescún* (Boyacá) in order to install some dismal stockades for the breeding of fighting bulls.

The alterations carried out with the intention of "improving" or decorating the interior living spaces or the outer appearance of colonial hacienda houses are other examples of the cultural and aesthetic mistakes characteristic of the Colombian monied classes. The modest colonial eaves have been replaced by cement projections which pretend to be supported by wooden corbels that are completely decorative and have highly pretentious outlines. Unwonted structural or decorative elements borrowed from religious architecture have arisen, like ghosts from previous epochs, at times in the form of a comical hotchpotch of columns, corbels, tie-beams and "Sevillan lazo" in the middle of halls and bedrooms. We shouldn't forget that ours is the epoch of the expository-bar, the altarpiece-dressing table, the oratory-study, the stable-gymnasium, the barn-tractor repair shop. A decidely modern age.

One exceptional case is the extreme modernization (1977-1979) of the colonial house of *Cortés* (Bojacá, Cundinamarca), which had previously been "republicanized" at the beginning of the twentieth century. There, for better or worse, the exterior volumetry and the spatial scheme of the existing are preserved, but the resulting architectural language is an uncomfortable mixture of the contemporary and the neo-colonial. There is certainly no room for doubt about the impression made by the present architecture of the house of *Cortés.*

Perhaps as a nostalgic or traditionalist reaction against the eclecticism of the end of the nineteenth century and beginning of the twentieth some hacienda or farm houses were built in Antioquia, Cauca and Cauca Valley which are the exact opposite of the modernization seen in *Cortés.* As though the past, together with the constructed forms associated with it, were recoverable, the simple incorporation of fake-antique "details" gave way, in some cases, to the complete building of rural houses with the exclusive use of late eighteenth century techniques and materials, a decision which lowered what might have been evocative or allusive to the level of a banal mimicry or imitation. The most outstanding example of this highly curious genre is the house of *Belalcázar,* on the outskirts of Popayán, that was finished in 1914! This "summary" of other colonial houses of Cauca, which *are* authentic, we must point out, does not lack formal and atmospheric grace, but it deceives even the most trained observer about its age. It is, in a certain way, the summit of the Colombian neo-colonial movement.

It is worth asking whether it might not be better to replace a badly updated traditional hacienda house with a totally modern version of the same. Such architectural creatures already exist in the country, done in a contemporary language which ranges from successful formal and atmospheric recreations by good designers to inexpressibly bad monstrosities intended for narcotics-traffickers and other personages of this kind. We may assume that, in the decades to come, an evaluation will be made of the buildings left by our epoch in the countryside and we will eventually be able to appreciate, through contrasts and similarities, the extraordinary quality and nobility of colonial haciendas and the ingenuous but attractive virtues of the republican period.

Opposite page, *El Molino,* Cogua, Cundinamarca.

Previous pages, left, above and below, and opposite page, *La Sierra,* Cauca Valley. The main face of the house, which was made symmetrical in the nineteenth century by means of the enclosure of the ends of the large gallery-balcony which commands a view of the valley and the addition of an axial path and entrance, breaking up of what should be a unitary space …an example of the transition from the colony to the republic. The notion behind the colombianization of hacienda houses was to superimpose eclectic forms upon colonial buildings, out of the conviction that all that was old was obsolescent. The best example is the house of *La Sierra,* Cauca Valley, also known as *El Paraíso,* the setting for the classic novel, *María,* and, due to its fame, now restored.

Above left, The old approach, on foot or horseback, to the house through the "back" patio, which conserves its perimetral walls but has been restored as a "formal" garden.

Below, The space of the gallery towards the view of the valley.

Opposite page, auxiliary corridors of the house.

Left and opposite page, *La Sierra,* Cauca Valley.

Left, The functionalism of the kitchen from the colonial period.

Right, The axis of symmetry imposed on the house as a sign of a new manner of spatial ordering. Note the continuous use of hexagonal bricks in the floor, a detail which originated in Andalusian, that is, Islamic traditions.

Above and right, *El Rabanal,* Ubaté, Cundinamarca. The architectural transition of this house did not consist of academic impositions but technical alterations which reversed the austere geometry of colonial building. The volume of the house remained the same, but the wooden columns of the upper floor gallery were multiplied and this additional weight made it necessary to "reinforce" the ground floor. The improvised stone columns show a complete indifference to (or ignorance of) the structural functioning of the old house but they turn out to be picturesque. Where the colony knew exactly what it was doing, the republic improvised.

Left and above, *El Salitre,* Sopó, Cundinamarca. A house admirably sitated in the landscape of the Cundinamarca high plain, it is a summary of the history of the rural architecture of the region. The colonial house which exists there was a late construction, done in the first years of the nineteenth century and replacing another, more modest one, whose date is unknown. In the second half of the nineteenth century the house acquired its present volume and area, together with the symmetrical treatment of the main façade, interior ceilings and a structural and spatial treatment of the interior patios characteristic of the urban architecture of Bogotá. It was modernized in recent years, creating window openings of a larger size and remodelling the the house with the cold geometrical precision characteristic of contemporary architecture.

Left and opposite page, *El Salitre,* Sopó, Cundinamarca. Note the presence of the inevitable axis of symmetry in the centre of the house and, on the right, the spatial treatment given to the small interior patio, with its considerable heights. It is worth comparing this space with that of the patio of the house of *Altamira,* Tenjo, Cundinamarca, in order to evaluate the changes in rural architecture between the colony and the republic.

Right, *El Hato,* Florida, Cauca Valley. After the superposition of eclectic elements upon the buildings of the colonial epoch, the next step in the transition between colonial hacienda houses and republican ones was the combination of Spanish constructive traditions with new compositional modalities to create new kinds of structures. A notable example of this process is the house of *El Hato,* whose airy volumetry establishes a different kind of relationship between house and place. Its rigorously symmetrical façades, built on the basis of freehand drawing, would allow it to be situated indiscriminately in any part of the region (which cannot be said, for example, of *La Concepción de Amaime* or *La Merced*). To a certain extent, *El Hato* inaugurates the epoch of the *modern* Cauca Valley hacienda house, one that is abstract and independent of its own siting, though without denying, for this reason, its colonial past. Note the very original combination, in this and other houses of the epoch, of peasant-style tile roofs and, below them, academic façades.

Previous pages, left and above, *Garciabajo,* Caloto, Cauca.

In its present form it makes up an architectural complex composed of eight different buildings which run from the eighteenth century to the last decades of the twentieth: the landlord's house, peons' quarters, storage sheds, corrals, stables, chapels, former annexes of the sugar mill. The main house itself, in contrast with *La Industria,* is a singular example of inspired improvisation and the taking advantage of existing ruins. In one of the many episodes of the civil war of 1884-1885 the sugar cane shed of the old sugar mill was burned down. The solid brick arcades of the building (from the eighteenth century) resisted the destruction of its roof. Upon the curious and enormous "hypostyle hall" formed by the old arcades a highly original new house was built, whose façade cannot avoid displaying the traditional look of a large gallery-balcony facing the landscape and terminated, at both ends, by closed volumes. A recent remodelling gave a certain decorative (and somewhat phantasmagoric) touch to what, in a bygone epoch, would have been strictly utilitarian features, like the great longitudinal extension of the arcades and mill's "fall of water".

Following pages, *La Industria,* Florida, Cauca Valley.

The final step in the transformation of colonial into republican architecture was to build entirely new structures. *La Industria,* according to B. Barney and F. Ramírez, "was designed by señor Francisco I. Caldas and built by the master of works Ramón Calero between 1917 and 1920". The house displays a rigorously symmetrical central façade, a now definitely *architectural* feature, which contrasts with the informality of the façades of houses like *La Concepción de Amaime,* in el Cerrito. What is gained in elegance of academic design is lost in atmospheric charm and relation to place.

Left above and below, and opposite page, *Santa Bárbara,* Tibitó, Cundinamarca. When an old hacienda house is modernized it frequently suffers from a decisive alteration of one of its most vital characteristics, the small original windows, which have been replaced by large panoramic ones that contradict the manner of looking at the landscape that existed in former epochs. It is worth thinking about what has been lost by gaining a broader view.

Opposite page, The house conserves the setting of a colonial interior patio, with its delightful atmosphere, dominated by a fountain which is beautiful but incongruous with the space which harbours it.

Previous pages, *Santa Bárbara,* Tibitó, Cundinamarca. Partly built during the eighteenth century and extensively reformed in the second half of the nineteenth century. These alterations were complemented by a recent modernization. The result is environmentally pleasing but devoid of the formal harmony characteristic of a building which is *exclusively* colonial. The addition of small enclosed balconies taken from urban styles and the enclosures in the façade of the front gallery of the upper floor (a space that was made to be open) are "modern" contributions.

Left above and below, and opposite page, *Buenavista,* Cota, Cundinamarca. The hacienda of this name was formed in the nineteenth century as a result of the dismembering of the extensive property of *Tibabuyes.* The house shown here is not a colonial structure but an entirely new building erected by D. José Ma. Urdaneta, its first owner, starting in 1872. Were it not for an unfortunate interior modernization done in 1967 and 1969, *Buenavista* would serve as an invaluable historical document and an extraordinarily interesting summary of the virtues and defects of the architecture of the republican period, among them, murals and other European-type decorations added by its second owner, Alberto Urdaneta. It still conserves the volumetry of its tiled roofs and exterior setting, including the coach yard in front of the house and its great axial entrance (right), in the manner of a French country villa, details that were added by Alberto Urdaneta from 1881 onwards.

Left, opposite page above and below, *Casa de los Mayores,* Titiribí, Antioquia. It was begun during the second half of the eighteenth century as part of the opening of mining territories in Antioquia. The house was enlarged and reformed on repeated occasions during the end of the nineteenth and parts of the twentieth centuries. A free-standing chapel arose alongside the house, together with a mini-village on the grounds of the hacienda. Note the similarity between these details and those which occurred in Guacarí, in the Cauca Valley.

Left, The use of very long eaves resting on struts became popular in Antioquia at the end of the nineteenth century.

Opposite page above and below, and right, *La Esmeralda,* Quindío. By way of comparison, we include this coffee-farm house which stands in a region settled during the Antioquian colonization of the centre of the country. Note the use of regional timbers, applied to the ceilings, floors and plinths, which was characteristic of artisanal construction in the second half of the nineteenth century in the regions of Quindío and Caldas. The tiled roofs that are seen in coffee-growing regions are generally smaller and less inclined than those of the hacienda houses of the colonial period.

Left, An example of the free adaptation, at the end of the nineteenth century, of such colonial traditions as interior patios and the establishing of a relation between the interior and exterior by means of continuous galleries. Due to the artisanal practices of the region and local historical circumstances, the atmospheric effects that have been achieved are notably different from those seen, for example, in the houses of colonial haciendas in the Cauca Valley.

Left above and below, and opposite page, *San Felipe,* Mariquita, Tolima.
Built (or rebuilt) during the second half of the nineteenth century and enlarged in the twentieth. One observes the same symptoms of the republican and even our contemporary epoch in a building where, as a kind of concession to the extreme tropical climate, the galleries are predominant.

Right, An example of the interesting appearance of an architectural volume that is new to this genre, which is the high open hall, that is, one that is devoid of bearing walls, for the same functional and climactic reasons. This would be suitable to modern rural architecture, but as far as colonial building is concerned, it does not take into account the architectural characteristics traditional to these climactic conditions in different regions of Nueva Granada.

Previous pages, opposite page and right above and below, *Cartama,* outskirts of Guamo, Tolima.
We may assign a mixed or transitional character to this house as well, analogous to that of the haciendas of the Cauca Valley at the beginning of the nineteenth century, without the new originality of the examples from the strictly republican period. The present appearance of the house of *Cartama* is due to recent reforms and a considerable modernization of what was a ramshackle house erected at the end of the nineteenth century on top of what must have been a modest eighteenth century house in the terrain of a hacienda founded at the beginning of the seventeenth century. It has conserved, throughout its history, the organization of the floor plan of the house in a longitudinal section, although the finished perimetral galleries with wooden pillars and lintels that it had in the nineteen-fifties no longer exist, having been replaced by "modern" colonnades. The enhanced volume of one of the ends, in the manner of Cauca houses, dates back to an undetermined epoch.

Left and above, *San Rafael,* Sesquilé, Cundinamarca.
This house, built at the end of the nineteenth century, is situated near that of *Chaleche,* a somewhat older house which is illustrated in the chapter entitled "Hacienda and House", and arose out of one of the subdivisions, in the republican epoch, of the hacienda of that name. Of course, its attractive volumetry is a direct representation of that of *Casablanca,* Sopó, and other colonial houses on the savannah of Bogotá, since there would have been no reason to "invent" a different architecture to confront the landscape of the region or to simply install a good rural house on a hillock that leans against the neighbouring mountains.

Above, The period in which the house was built is deduced from the use of the roof frames in "Roman" wooden trusses with a king post, which were the only features of traditional methods remembered by nineteenth century builders once the tradition of the Santa Fe version of "par y nudillo" had been abandoned or forgotten. The large modern windows decisively alter the interior spaces.

Left above and below, and opposite page, *La Esmeralda,* El Cerrito, Cauca Valley. It was inadvertently included in *Casa Colonial.* In fact, we are dealing with a wholly republican hacienda house, since its building dates back to the last years of the nineteenth century, with reforms and additions in the second decade of the twentieth. It seems that a small hut existed on the site in a previous but undated epoch. The house includes traditional compositionl elements present in colonial building, such as the peripheral galleries with wooden pillars and lintels and the elevation of a small section in the upper floor, but all this is combined with flat ceilings and a certain dimensional simplification and aesthetic impairment of the systems of columns, railings, grilles, doors and windows, which are formal indications of the republican period. The house continues the tradition of an inspired choice of place and siting, taking advantage of a hill that dominates the surrounding area.

Left and above, *La Unión,* Armero, Tolima.
The hacienda houses of the region are, in general, later than the foundation, in the middle of the nineteenth century, of the city of this name, which leads us to suppose that it was built from 1875 onwards. There is a typological similarity between *San Felipe,* in Mariquita, and *La Unión,* given that they are adapted in a similar way to the tropical climate of the region. The preponderance of ample galleries and the location of the halls of the upper floor are architectural elements that were implanted on rural construction in the central regions of the country during the republican epoch.

Left, and opposite page above and below, *Juan Blanco,* Santa Fe de Antioquia.
This, like most hacienda houses of the region, had its original sections built during the height of the colonization and exploitation of Antioquian territory which started at the beginning of the eighteenth century.

Left, The distinctive architectural feature of the hacienda houses of Antioquia continues to be their perimetral galleries on wooden columns resting on masonry bases. The spatial emphasis in this and other houses of the region is not on the interior patios but towards the exterior faces of the building.

Right, The abundant vegetation around the house has been conserved as a basic element in the relation between the house and the place. The environmental quality thus achieved adds authenticity to both parts.

Left and opposite page, *Mesitas de Santa Inés,* Cachipay, Cundinamarca. Although subjected to a highly extensive reconstruction, the house conserves a good part of the spatial ordering characteristic of urban residential buildings at the end of the nineteenth century in Bogotá, in which the corridors and galleries compose the largest area. It is possible that the house arose in the period between 1870-1900, when the railway between Bogotá and Girardot stimulated the socio-economic development of the region and Cachipay was one of the important stations along the route. Note the traditional use of galleries built with wooden pillars and lintels, combined with floors, ceilings and partitions in the same material.

Above and right, *Santa Rosa,* El Rosal, Cundinamarca. An outstanding example of a farm house of the republican epoch situated in a property that resulted from a subdivision of subdivisions of lands of the colonial epoch that took place during the course of the nineteenth and twentieth centuries. *Santa Rosa* replaced, in the first years of the twentieth century, a modest hut that stood on the site. The case shows successive enlargements around a central space and the correct use of traditional elements like roof tiles and perimetral galleries.

Left and opposite page, *Santa Rosa,* El Rosal, Cundinamarca.

Left, The age of the house of Santa Rosa is nearly the same as that of the wax palm planted in its central patio.

Right, In the interior of the sitting room there is a reform which unusually contradicts what is usually found in a house of the republican period or an older one that has been strongly "republicanized", namely, the elimination of flat ceilings intended to hide the rustic roof frames.

Above and right, *Los Arboles,* Madrid, Cundinamarca. The hacienda of this name was established at the end of the eighteenth century as part of the process of breaking up large properties in the southwest of the savannah of Bogotá. The structure of the house and its spatial ordering around a beautifully-proportioned central patio of the colonial period are included in the present chapter because of the interesting and well-conceived process of integral republicanization carried out between the last decades of the nineteenth century and the first two decades of the twentieth.

Opposite page and right, *Los Arboles,* Madrid, Cundinamarca. A demonstration of how it is, in fact, possible for an eighteenth century house to pass through the Europeanizing eclecticism of the nineteenth century and later undergo a twentieth century orientation without jeopardizing its original architecture or losing its intrinsic value. An historical document.

Left above and below, and opposite page. *Alcalá,* Samacá, Boyacá.
The earliest section of the house dates from the eighteenth century but it was enlarged and altered during the nineteenth. It is, like *Villa del Rosario,* a borderline case, in which the original building has lost its authentic architectural qualities through successive transformations but still preserves some spatial and environmental traits. Note the modern use of a republican finishing wallpaper.

Opposite page, right above and below, *Aposentos,* Cogua, Cundinamarca. Like *Los Arboles,* this house is from the colonial epoch, of which it conserves the volumetry and a good part of its surrounding walls, which means that it maintains its environmental relation with the place where it stands. Its inclusion among hacienda houses of the republican period is due to the same reason invoked for the previous example: the abundance and quality of electic elements of the nineteenth century, together with the decoration and the interior and exterior spaces of the house.

Opposite page, Note the singular treatment of the front garden, with a grating characteristic of an old-fashioned estate house in Bogotá.

Right below, The variegated collection of colonial gold-leaf work in the republican oratory.

Opposite page, and right above and below, *El Noviciado,* Cota, Cundinamarca. It was founded in the first third of the seventeenth century by the Jesuit priest José Hurtado. It is not certain whether the building which exists today, which is now the country seat of the University of the Andes, is the same as the one referred to in eighteenth century documents, or, as technical considerations indicate, it is a partial reconstruction, carried out in the nineteenth century, of the ruins of the building that was abandoned by the Jesuits when they were expelled from Nueva Granada at the end of the eighteenth century. *El Noviciado* is included here because of the republican-period aspect and technical features which it now shows.

Opposite page, The gallery of the ground floor which shows the stone columns, taken from the priest's house of Chia, which recently replaced the older wooden pillars and the brick floor that replaced the adobe one.

Right, above and below, Façades of the house which allow one to see its truncated or incomplete character.

Left, *El Noviciado,* Cota, Cundinamarca. In the early years of the twentieth century the upper gallery of *El Noviciado* was enclosed in the usual republican manner, that is, with enclosures formed by wooden parapets and windows with small divisions. The recent restoration modernized those elements. The stone columns of the lower gallery, although they are elements from the colonial epoch, represent a modern intrusion, due to their incongruity. Note, on the right of the photo, the addition of the caretaker's house, which is devoid of any relation with the main building.

Above and right, *El Vergel,* Ibagué, Tolima.
The observations made about the houses of *San Felipe,* in Mariquita, and *La Unión,* in Armero, apply here, since they all belong to the same geographical region and share a similar architectural nature.
El Vergel has been notably modernized, but it conserves the atmosphere and dimensions of its second storey galleries.

Above and right, and following page, *La Ramada,* Madrid, Cundinamarca.
The house is the result of the successive dividing and restructuring of haciendas on the savannah of Bogotá between the second half of the eighteenth century and the first half of the nineteenth. It appears that only a part of the present house dates from the colonial period, the remainder being the product of enlargements and reforms that extended into the first decades of this century. Finally, the house was subjected to a modernization needed to adapt it to contemporary styles of life.

Opposite page, The enclosure, through latticed windows, fretwork and parapets, of corridors and galleries that were originally open, is an architectural process common to all hacienda houses, no matter of what epoch, which stand in cold or temperate climates. It is assumed that greater comfort and a more elegant appearance will be achieved in this way.

Opposite page and right, *La Ramada,* Madrid, Cundinamarca. The house conserves a great part of its colonial roof frames, to which have been added modern elements which did not exist in in the hacienda houses of that epoch but were common in the republic, such as the enclosure of the old perimetral corridors or the furniture incrusted in the walls.

Left above and below, and opposite page, *Casa de Teja,* Madrid, Cundinamarca.
A house, no longer of a hacienda, but a farm, situated in a terrain which arose from the third or fourth breaking-up of the large properties on the western part of the savannah of Bogotá. This is a completely republican-period building which dates from the end of the nineteenth century. Observe the preponderance of glassed-in enclosures. The transparency thus achieved is an architectural trait characteristic of the republican and modern epochs and represents the extreme environmental opposite of the long walls, with small openings, of the colonial period.

Opposite page and right, *La Iraca,* Tibabosa, Boyacá. Also known, several decades ago, as *La Esperanza*. Only one part of the longitudinal section belonging to the house built at the end of the seventeenth century still exists. Successive alterations carried out in the nineteenth and twentieth century considerably altered its volumetry and exterior spaces. Nowadays the house has a completely republican or modern appearance, the only genuinely antique elements being the walls and surrounding spaces.

Left, High gallery of the wings added to the colonial house in the nineteenth century.

Left, and opposite page, *La Iraca,* Tibabosa, Boyacá. A traditional element conserved and reconstructed numerous times in *La Iraca:* the country oven.

Opposite page, The axis of symmetry and main access path imposed on the house during the nineteenth century.

Left above and below, and opposite page, *La Julia,* Buga, Cauca Valley.
The most notable remodelling in the western part of the country of an eighteenth century colonial house, carried out in the second decade of the twentieth century. The author of the work, which is highly interior and fully respectful of the original volumetry and spatial organization, was Alfredo del Diestro.

Above, The inevitable axis of symmetry on the front of the house.

Below, The decorative epidermic treatment of the perimetral galleries.

Opposite page, The interior treatment of the house included faceted corners in the reception rooms, mouldings in plasterwork, ceilings and cornices in brass and wooden caissons, wallpaper, French parquet floors and mural painting done on canvas and wood. Such a decorative display is not only surprising but unique for a country house of that region.

Left above and below, and opposite page, *El Corzo.* Facatativá, Cundinamarca. The Alberto Angel Montoya House. The colonial property of this name has existed since the first third of the seventeenth century but the present house is the result of the integral interior treatment of the spaces from the colonial epoch with finishings and decorations which date from 1881 and 1885 and were applied by the same builders who achieved a completely similar effect in the upper storey of the house of the Germán-Ribón family in Bogotá, which is now the Museum of the Nineteenth Century (run by the Coffee-Growers Cultural Fund).

Left above, As in the example just cited, the precints of the colonial epoch were joined together by means of French-made folding doors, thus creating a decisive spatial change.

Left below, The house conserves part of the volumetry of the colonial epoch.

Left and above, *Casa Meridor,* Tenjo, Cundinamarca. This is not a celebrated or imposing hacienda house but the modest seat of a farm in the Valley of Tenjo. It is, in the same manner, the fruit of a recent modernization done to a building that cannot be much older than a hundred years, that is, a farm house of the republican period. There is, without necessarily striking a discordant note, a certain logic to choosing it to close the subject of architecture in the Colombian countryside, because the renovation of the house makes a discreet use of traditional elements and spatial organization and shows a certain restraint in the employment of large windows and other contemporary features. In short, the problem of design implicit in this case was resolved with simplicity, without pretending to add an architectural message. What is debateable about it is much less important than its fundamental character: the conservation of a splendid siting on a hillside and a sustained environmental relation between the house and the surrounding landscape. This, which appears simple and obvious, is much more than we may say about most of the buildings which make up what might be called the history of rural architecture in the country.

Left and above, *Boitá,* Sesquilé, Cundinamarca. It conserves a very dramatic siting, on the top of a hill which dominates a good part of the surrounding landscape. The distinctive architectural feature of the house, which was built at the beginning of the eighteenth century on lands which were part of the extensive land-grant property – and later hacienda – of *Hatogrande,* are the two great pinnacles, characteristic of some religious building (where they would not be so strange) and used in the house in the manner of decorated counterweights placed above the points most highly stressed by the diagonal thrusts of the roofs upon the bearing walls. The colonial volumetry and and central patio of *Boitá* went through a discreet *republicanization* at the end of the nineteenth century, when some points of the exterior galleries were partly shut in and enclosed. In the nineteen-fifties the Swiss architect Víctor Schmid applied his singular eclecticism to the house. Later on, the house was altered again in order to install panoramic windows, ceilings, etc. and partly modify the arrangement of small spaces favoured by Schmid. The beautiful presence of the old tiled roofs survives, with their strange pinnacles that are now accompanied by modern chimneys.

Left above and below, and opposite page, *Villa del Rosario,* Samacá, Boyacá. A former residence of the Jesuits up to the middle of the eighteenth century, which explains the annexed "doctrinary" chapel. The recent history of the house witnessed a modernizing alteration in the early nineteen-eighties, which included the implanting of a stone portal (that was not original) on the façade that radically changed what had been a harmonious volumetry, and the ultra-modern enclosure of the galleries of the upper storey.

Left and above, *Fute,* Soacha, Capital District. This was one of the oldest haciendas of the savannah of Bogotá, built at the beginning of the seventeenth century. In the eighteenth it became the property of the Jesuits and they were probably responsible for the construction (or remodelling) of the original house. In the nineteen seventies *Fute* was subjected to a considerable reconstruction and modernization.

Opposite page above and below, and right, *Fute,* Soacha, Capital District.

Opposite page below, In the colonial period architectural elements were created in order to look out, from them, onto the outside world. Nowadays they are created with the exclusive aim of being seen, preferably for a brief moment, from the outside. Hacienda houses have changed from being utilitarian tools to serving as stage sets.

Right, Of its colonial past, only a part of the general volumetry of *Fute* remains. The modernization that it underwent was just as afraid of meeting the challenge of starting from nothing to construct something new in a contemporary language as it was of undertaking a restoration in the strict sense of the word, clearly oriented towards rescuing what was original.

Above and right, *Cortés,* Bojacá, Cundinamarca.
It is possible that a modest house, which was gradually enlarged over the centuries, existed in the hacienda of this name, at the beginning of the eighteenth century. Documentary evidence shows the great expansion and subsequent dividing of *Cortés.* In the middle of the nineteenth century one of its owners "rebuilt" the colonial house, which, means that he practically replaced it. This leads us to assume that the house which survived into the twentieth century was a purely republican one. In the nineteen-seventies, the house of *Cortés,* was subjected to a radical modernization, not restoration. This approach is perhaps more authentic than in other houses. There is no doubt, in *Cortés,* about what we are dealing with. It is a modern house that, from a certain distance, evokes the volumetry and spatiality of the republican building that preceded it on the site, even though it conserves, in modern dress, some walls from the latter. A certain discretion and moderation of design is seen in the present appearance of *Cortés,* which is devoid of decorative posturing and allows its volumetry and spatial organization to be seen.

Above, Interior of the south section of the house, which shows the reception rooms obtained through the elimination of tranversal walls and the superficial evocation, by means of neutral surfaces, of the space determined by the old roof frames in rustic timber.

Left and opposite page, *Casablanca,* Madrid, Cundinamarca.
A house built at the end of the seventeenth or the beginning of the eighteenth centuries existed in the place, of which there only remains the platform on which the present L-shaped sections, devoid of perimetral galleries, now stand. During the second half of the nineteenth century, the house underwent a number of alterations, one of which tended to make the general volumetry symmetrical, creating a second lateral wing on the front in the form of a U and opening the inevitable axial entrance. The present appearance of the house derives from alterations done in the nineteen-forties. The original walls and roofs have disappeared. The irreplaceable environmental charm of the colonial or the eclecticism of the republican are also absent. The house displays, by contrast an uncertain modernity, whose siting and volumes are correct.

Above and right, *Hatogrande,* Sopó, Cundinamarca. Current country residence of the president of Colombia. The house is similar to *Buenavista,* illustrated in preceding pages. Like the latter, the house sold by a wealthy landlord to the Colombian government in the nineteen-forties did not have, as is generally believed, a colonial origin. The seventeenth century house of the same name was situated at a distance of one and a half kilometres from the one which now bears the name of the old land-grant estate of *Hatogrande.* There was no house on the present site until the period 1858-1861, when a gentlemen by the name of Silva built the first sections. Only traces of the foundations and some walll stumps remain of that building. They are buried beneath some fourteen separate stages of construction, demolition, reconstruction, enlargement, decoration and redecoration carried out, since 1947, according to the whims of almost all the presidents who have served since that date. The result of so many clashing and contradictory architectural and decorative interventions is a building that cannot be classified as anything other than a completely modern one.

Conservation and remembrance of hacienda houses

"The originality of a work sometimes depends
on what its author cannot do.
There is a creative impotence."
NICOLAS GOMEZ DAVILA.
Glosses on an implicit text.

The content of the illustrations which precede this chapter refer to the process whereby hacienda houses were transformed when the republican era arrived and to the new rural architecture that was characteristic of that period. At the end of the previous chapter we assembled some examples of modernizations of hacienda houses that, being originally colonial or republican, suddenly confronted the twentieth century, not always with fortunate results. We have already seen how a formal and atmospheric continuity between the architecture of the Spanish colony and the republic was feasible, up to a point, and took place in favourable circumstances, but was then nullified by the absence of an historical and cultural harmony between those epochs and the present one. The examples illustrated in previous chapters confirm the impracticality of this conversation among the deaf, that is, between what existed before and what was added in the recent epoch. They are all cases of a radical architectural intervention which, at times, pretended, in a very superficial way, to take into consideration the past existence of constructed forms and, at times, adopted an outright avant garde attitude that was indifferent to any architecture from other periods. The reader will have to draw his own conclusions about the critical evaluation of what is shown here, with the idea of creating a pause for reflection about such matters. Is this the path we should follow, so that hacienda houses continue to exist? Is modernization a more authentic and worthwhile process than the imitation of the architecture of the past? Is it better to build a hacienda house for the present period than continue to defigure and repair one built centuries ago? This essay does not pretend to give definitive answers to such qestions, but rather to provoke our readers to debate the question for themselves.

While it would be possible to list the Colombian hacienda houses from the colonial or republican periods which have disappeared or been disfigured, semi-destroyed or abandoned, it is a long and unrewarding task. Rural architecture has always been less important than urban building, both in the popular consciousness and in the cultural media. A proof of this is the fact that historians, critics and restorers only concern themselves with peasant or hacienda houses when the subject of urban architecture, whether monumental or not, has been exhausted. Even during the Spanish colonial period, the country house was always the "second" or "other" residence, that which could be used – and abused – because it was felt to have the status of an everyday object, in contrast to the importance given to a family seat in the city. The hacienda house was created within a socio-economic system in which it was an *essential* element. When historical circumstances changed and a capitalism that was ever more consumer-minded arose in the nineteenth century, the hacienda house acquired a commercial value as an article of real estate, a meaning as a symbol of a social class and functions as a weekend house or social seat. The significance of the country house diminished to the same extent that the lands around it became more valuable being thus relegated to a *secondary* status. It is possible, in the twentieth century, to own and work small or large rural properties without necessarily building or maintaining a house on them. A prefabricated shack or a watchman's hut is sufficient. Nowadays the house is not secondary but sumptuary.

Today, in Colombia, a colonial hacienda house is a splendid luxury and a beautiful adornment to the landscape, a delight for the senses and the mind, as well as a cultural spectacle of the first order, but it would be fastidiously elitist to invoke these considerations as a reason to support the conservation of such a singular genre of historical architecture. The conservation of a heritage of rural architecture is an enterprise that more properly belongs to countries like England, with its celebrated family seats, or the United States, with its great historical haciendas of the Mid - and Far-West, Southern cotton and tobacco plantations and many more, all of them considered to be part of the national heritage and main-

Opposite page, *Hatogrande,* Sopó, Cundinamarca.

tained as residences, tourist attractions or as a matter of national pride and rejoicing.

The essentially utilitarian nature of the hacienda house is a vital contribution to the relation between its owners or residents and the constructed forms of the countryside. There are many reasons that play a role in the decision to discard, from everyday life, a farming tool, draught animal or house that is already old, obsolete, worn out or useless. In present-day circumstances, to pretend, for simple cultural reasons, that we must (or can indefinitely) prolong the material existence of a hacienda house that is besieged by urban development, deprived of its surrounding lands, damaged by the passage of time or devoid of its vital reason for being, not only constitutes a bitter self-delusion but is also impractical, except for a highly limited number of owners possessed by the demon of ancestral pride or an addiction to the past, myth and legend and perhaps, in a very few cases bordering on madness, by the poetry of the place and the magical powers of architecture.

It is not legally possible to oblige a landlord to maintain a sentimental relationship with his country house. Nor can you calculate a hard-currency price per hectare of landscape, with a surcharge when the view is "marvellous" instead of just "pretty" or a discount when nature is invaded by a cultivation of flowers covered by a shroud of plastic. The hacienda house as a collector's piece that is perpetuated by nostalgia or snobbery would no longer be a living thing but an architecture whose fate is embalmed.

An ancient hacienda house does not deserve being placed alongside your great-grandmother's wedding dress or a daguerrotype of the ancestor who fought in God only knows what civil war of the nineteenth century. To use it as a hotel, a weekend house, a "convention centre", a picturesque restaurant a "spiritual (?) retreat", the headquarters of a sect, a gymnasium for aerobics or an exclusive country club places it on another planet, even when these activities take place in what were once hacienda houses. A half-utilitarian, half-emotional justification is usually invoked for such motley adaptations, namely, that the old hacienda house provides a physical and environmental framework that has a certain grace and attractiveness which are absolutely lacking in all the examples of contemporary architecture that have been specifically created to deal with these modern uses. The true dimensions of the "creative impotence" characteristic of our age can be seen in this chronic inability to provide something that can compete in formal and atmospheric quality with the humble structures built in the past throughout the length and breadth of the Colombian countryside.

There, no doubt, is part of the cultural heritage of the country, in largely squandered and only affectionately cared for here and there and as an exception rather than a rule. Nor we should we forget, of course, that an inheritance always comes after a death and does not necessarily imply a committment to the deceased or a resurrection. The colonial hacienda house of the savannah of Bogotá that arises from the dawn mist surrounded by an ocean of plastic sheets, dirty with industrial effluents, which cover vast cultivations of flowers is a horrible accusation against the present age, but it is not a great example of how to conserve the architectural heritage of the Colombian countryside. Nor is the eighteenth century rural house dressed in a dreadful shroud of cement dust, lime and residues of coal and iron in the Sogamoso Valley, in Boyacá. Even less so are the stumps of the walls of a beautiful sugar-mill hacienda house in the Cauca Valley, which was first abandoned and then intentionally burned down in order to erase it from the physical and cultural map of the country and the collective memory.

More and more hacienda houses, including *Santillana* and *El Alisal* now only live on in our memory, in books, photographs and documents which prove that they were entitled to, at least, an honourable burial. Look at the care and affection behind the photographs of the hacienda houses included in this book, which avoid showing their damages, scars, vandalism and destruction. These pictures are beautiful but one-sided, with an intentional and subjective inclination towards idealizing an architectural genre which, only a short time ago, came to know ugliness and foolishness.

The hacienda house was not created to be an untouchable monument but as an instrument of work and shelter. The colonial hacienda house no longer exists in the terms in which it was created, nor does the hacienda itself.

The utilitarian and functional reasons which gave them vitality and a reason for being disappeared a long time ago.

In the light of this consideration, the somewhat arbitrary attempts, by the Council of National Monuments, to grant protection to a number of colonial hacienda houses may seem exotic or irrelevant. Unless it is accompanied by expensive maintenance grants, the effort to conserve such houses, which often is due to extra-architectural reasons, is meaningless. A few hacienda houses still carry out their original functions. Many others, having already been mummified, now have "compatible uses", which is a polite way of saying that they are dedicated to the culturally acceptable activities described above. The destruction or abandonment of colonial hacienda or farm houses has reduced their surviving numbers to less than a quarter part of the total which we can reasonably assume to have been originally built in Neo-Granadian or Colombian territory. If we take into account that they were never built with the idea of being eternal, but only moderately durable, the problem of their conservation becomes even more difficult. From the beginning, the rural buildings of Nueva Granada required a large measure of work, patience and affection from their owners or inhabitants in order to maintain them in a reasonable state. At times, what some proud owner boasts of as being the "seventeenth century" house that now stands in a landscape invaded by quarries, billboards, weekend "chalets" and "regional cuisine" restaurants is, in reality, the third version of the one which was original to the site, and was only built, in the late nineteenth century, with peasant stubbornness, after the first burnt down from a lightning bolt and the second was demolished once it fell into disrepair.

Such a singular history, which was possible in the past, appears, now that we are in the last years of the twentieth century, to have reached its end. The historical stage of nostalgia and memories has now arrived, with its sentimental idea of restoration and the cult of the antique, with dirt and discomfort included. The hacienda house will take its place alongside wind-powered galleons, the ox-drawn plough, gilded carriages and medieval Spanish.

Hundreds of farm and hacienda houses exist in the country, some of them incredibly beautiful and useful, others less so, which make up a vast but highly threatened cultural heritage. For the person who would like to discover and appreciate this enormous wealth, the Colombian countryside is an unknown and largely unexplored territory. Even the most celebrated hacienda houses are only superficially known and understood. Myth, legend and cultural falsehoods cover and obscure them, as happens with the personages of the country's political history. It would be easier to capture the essential truth of the countryside architecture of the colony or republic if we remember, at each step, that a nation, throughout its history, produces exactly the architecture it deserves and no other.

It is certain that one cannot love or understand what one does not know. Why build a house in a beautiful country setting if you are not going to look after it as though it were a living being and lavish upon it the kindness and tolerance you dedicate to someone you love?

To enter an ancient hacienda house which has been mistreated by time but touched by the grace or magic of constructed forms is to enter into contact with oneself, with what we were, with what we would like to have been. The country house is a city house that has gone to paradise. As Antoine de St.Exupéry says, in *Citadelle:*

"Therefore find provisions on the horizon and bring home the beatitude that you have found. And the house transforms itself because some where the prairie, the dawn and the sea exist. Since everything opens onto something vaster than itself. Everything becomes a path, a route and a window opening onto something vaster than itself."

Footnotes

1. José Antonio Terán Bonilla. *Arquitectura Rural en México. Las Haciendas de un Región.* ("Rural Architecture in Mexico. The Haciendas of a Region"). In the volume *Estudios Sobre Arquitectura Iberoamericana.* ("Studies of Latin American Architecture"). Ed. Junta de Andalucía, Consejería de Cultura y Medio Ambiente, Sevilla. 1990.
2. *Diccionario de Autoridades.* ("Dictionary of Authorities"). Real Academia Española. Madrid. 1726.
3. Fernando Chueca Goitia. *Historia de la Arquitectura Española.* ("History of Spanish Architecture"). Volume I. Ed. Dossat. Madrid. 1965.
4. Chueca Goitia. Op. cit.
5. See the historical "gap" in the work cited of F. Chueca Goitia. In the chapters on the Gothic, rural architecture seems not to exist.
6. Antonio M. Bernal. Lecture given in the course on *Referentes Andaluces en la Arquitectura Iberoamericana.* ("Andalusian References in Latin American Architecture"). Baeza, España. 1988.
7. Chueca Goitia. Op. cit.
8. Germán Colmenares. *Cali: Terratenientes, Mineros y Comerciantes.* ("Cali: Landowners, Miners and Merchants"). Ed. Banco Popular. Bogotá. 1983. *Popayán, Una Sociedad Esclavista.* ("Popayán, A Slave-owning Society"). Volume II of *Historia Económica y Social de Colombia.* ("Economic and Social History of Colombia"). La Carreta Inéditos.Bogotá. 1979.
9. Benjamín Barney and Francisco Ramírez. *La Arquitectura de las Casas de Hacienda en el Valle del Alto Cauca.* ("The Architecture of the hacienda houses of the Upper Cauca Valley"). El Ancora Editores. Bogotá. 1994.
10. Antonio Bonet Correa. *Andalucía Barroca.* ("Baroque Andalusia"). Ed. Polígrafo. Barcelona. España. 1978.
11. Chueca Goitia. Op. cit.

11 A. Gisela von Wobeser. *La Formación de la Hacienda en la Epoca Colonial. El uso de la tierra y el agua.* ("The Formation of the Hacienda in the Colonial Epoch. The use of land and water"). Ed. Universidad Autónoma de México. 1989.

12. Margarita González. *El Resguardo en el Nuevo Reino de Granada.* ("The *Resguardo* in the New Kingdom of Granada").La Carreta Inéditos. Bogotá. 1979.

13 and 15. *Transcripción de las Ordenanzas de Descubrimiento, Nueva Población y Pacificación de las Indias dadas por Felipe II en 1573.* (Transcription of the Ordinances of Discovery, New Settlement and Pacification of the Indies issued by Philip II in 1573.) Ministerio de Vivienda, Madrid. España. 1979.

14. Carlos Martínez. *Apuntes sobre el Urbanismo en el Nuevo Reino de Granada"* ("Notes on Urbanism in the New Kingdom of Granada"). Ed. Banco de la República. Bogotá. 1964.
16. Ricardo Rendón Garcini. *Haciendas de México.* ("Haciendas of Mexico"). Ed. Fomento Cultural Banamex. México. 1994. Gisela von Wobeser. Op. cit. No. 11.
17. Margarita González. Op. cit.
18. Germán Colmenares. *Popayán, Una Sociedad Esclavista.* Bogotá. 1979.
19. Germán Colmenares. *Las Haciendas de los Jesuitas en el Nuevo Reino de Granada.* ("The Haciendas of the Jesuits in the New Kingdom of Granada"). Ed. Universidad Nacional de Colombia. Bogotá. 1969.
19. A.Ortega y Gasset, José. *Teoría de Andalucía.* ("Theory of Andalusia"). Complete Works.Vol.VI. Ed. revista de Occidente. Madrid. 1961.
20. R. Rendón Garcini. Op. cit.
21. G. Colmenares. Op. cit.
22. Source of data: Juan Carrasquilla. *Quintas y Estancias de Santa Fe y Bogotá.* ("Country Estates and Farms of Santa Fe and Bogotá"). Ed. Banco Popular. Bogotá. 1989.
23. G. Colmenares. *Cali: Terratenientes ...*
24. G. Colmenares. *Popayán ...*
25. See reference to the text of Maricruz Aguilar included in the summer course of the Antonio Machado University (Baeza, 1988) in Chapter I.
26. G. Colmenares. *Cali: Terratenientes...*
27. G. Colmenares. *Popayán ...*
28. G.Telléz. *Cultura y Tecnología en la Arquitectura Colonial de la Nueva Granada.* ("Culture and Technology in the Colonial Architecture of Nueva Granada"). Columbia University, New York. USA. 1987. McGill University, Montreal. Canadá. 1987.
29. See references to the participation of Petrés in the house of *Aposentos* (Simijaca, Cundinamarca) in *Casa Colonial* ("The Colonial House"). (G. Telléz. Villegas Editores. Bogotá. 1996). Chapter, *Creación y Orden* ("Creation and Order"). The main contribution of Petrés was, without doubt, the giving of an academic symmetry to the principal facade and the volumetry of the house.
30. The wooden framework of roof support in *par y nudillo* consists of two diagonal elements, the *pares* or *limas,* which rest on beams which run along the bearing walls. These beams are linked together by ties or crosspieces. The point of union of the *pares* is made rigid through a horizontal piece which forms a triangle, the *nudillo.* The variation, *par e hilera,* places the *pares* at a more acute angle and does not require the *nudillo.* The *hilera* is the part that runs through the roof saddle, the highest point of the framework. These systems are of Islamic origin, their structural principles being opposite to those of the *cercha* or Roman framework.
31. Camilo Pardo Umaña. *Haciendas de la Sabana de Bogotá.* ("Haciendas of the Savannah of Bogotá").Villegas Editores. Bogotá. 1988.
32. Eduardo Caballero Calderón. *Diario de Tipacoque.* ("Diary of Tipacoque"). Ed.Siglo XX. Bogotá. 1950.
33. Mario Sartor. *La Vivienda Mediterránea y la Tipología de la Casa Colonial Americana.* ("The Mediterranean Dwelling and the Typology of the American Colonial House"). In *Estudios sobre Arquitectura Iberoamericana.* Sevilla. España. 1990.
34. Germán Téllez. Chapter *La Casa de Hacienda* ("The Hacienda House") in *Historia del Arte Colombiano.* Ed. Salvat, Barcelona, España. 1975
35. Antoine de St. Exupéry. *Citadelle. Oeuvre Compléte.* Collection La Pléiade. Paris. 1947.
36. Orlando Fals Borda. *Historia doble de la Costa.* ("Double History of the Coast"). Vol. 1 *Mompox y Loba.* Carlos Valencia Editores. Bogotá. 1979.
37. Néstor Tobón. *La Arquitectura de la Colonización Antioqueña.* ("The Architecture of the Antioqueñan Colonization"). 5 vols. Villegas Editores. Bogotá. 1979-85.
38. See: Sylvia Arango. *Historia de la Arquitectura en Colombia.* ("History of Architecture in Colombia"). Ed. Universidad Nacional de Colombia. Bogotá. 1993.
39. In a wall plan which exists in the house of *Buenavista,* done in the second half of the nineteenth century, we find the location of the old (colonial) house of the same name and the "new" one, built around 1872, with a symmetrical ground plan, although not upon the walls of the first, as was said.

Opposite page, *Los Aposentos,* Chocontá, Cundinamarca.

Bibliography

AGUILAR, Maricruz. *Haciendas de Olivar.* ("Haciendas of Olivar"). Text published in an unidentified magazine and included as a fotocopied appendix to the course *Referentes Andaluces en la Arquitectura Iberoamericana* ("Andalusian References in Latin American Architecture"), held in Baeza, Andalucía, in 1988.

ALBERTI, Rafael. *Antología Poética.* Ed. Losada. Buenos Aires 1950.

ANCONA Riestra, Roberto. oordinator. Various authors. *Arquitectura de las Haciendas Henequeras.* ("Architecture of the Henequeran Haciendas"). Ed. Universidad Autónoma de Yucatán. Mérida. México. 1996.

ARANGO, Sylvia. *Historia de la Arquitectura en Colombia.* ("History of Architecture in Colombia"). Ed. Universidad Nacional de Colombia. Bogotá. 1993.

AUTORES VARIOS. (including M. Sartor, G. Téllez and A. Terán). *Estudios sobre Arquitectura Iberoamericana.* ("Studies of Latin American Architecture"). Consejería de Cultura y Medio Ambiente. Junta de Andalucía. Sevilla. 1990.

BARNEY, Benjamín y RAMIREZ, Francisco. *La Arquitectura de las Casas de Hacienda del valle del alto Cauca.* ("The Architecture of the Hacienda Houses of the Upper Cauca Valley"). El Ancora Editores. Bogotá. 1994.

BONET Correa, Antonio. *Andalucía Barroca.* ("Baroque Andalucia"). Polígrafo editores. Barcelona, España. 1978.

CABALLERO Calderón, Eduardo. *Diario de Tipacoque.* ("Diary of Tipacoque"). Ed. Siglo XX. Bogotá. 1950.

CAMBREZY, Luc y LASCURAIN, Bernal. *Crónicas de un Territorio Fraccionado. De la Hacienda al Ejido.* (Centro de Veracruz). ("Chronicles of a Broken-up Territory: From the Hacienda to the Common"). Ed. Larousse-ORSTOM. México, 1992.

CARRASQUILLA, Juan. *Quintas y Estancias de Santa Fe y Bogotá.* ("Country Estates and Farms of Santa Fe and Bogotá"). Ed. Banco Popular. Bogotá. 1989.

CHUECA GOITIA, Fernando. *Historia de la Arquitectura Española.* ("History of Spanish Architecture"). Vol. I. Ed. Dossat. Madrid. 1965.

COLMENARES, Germán. *Historia Económica y Social de Colombia.* ("Economic and Social History of Colombia"). Volume II. *Popayán una Sociedad Esclavista. 1680-1800.* ("Popayán: A Slave-owning Society. 1690-1800"). La Carreta Inéditos. Bogotá. 1979. *Sociedad y Economía en el Valle del Cauca.* ("Society and Economy in the Cauca Valley"). Volume I *Cali: Terratenientes, Mineros y Comerciantes. Siglo XVIII.* ("Cali: Landowners, Miners and Merchants. XVIII Century"). Ed. Banco Popular. Bogotá. 1983. *Las Haciendas de los Jesuitas en el Nuevo Reino de Granada.* ("The Haciendas of the Jesuits in the New Kingdom of Granada"). Universidad Nacional de Colombia. Bogotá. 1969.

DIAZ de Zuluaga, Zamira. *Sociedad y Economía en el Valle de Cauca.* ("Society and Economy in the Cauca Valley"). Volume II. *Guerra y Economía en las Haciendas. Popayán 1780-1830.* ("War and Economy in the Haciendas. Popayán 1780-1830").

FALS BORDA, Orlando. *Historia doble de la Costa.* ("Double History of the Coast"). Vol. 1 *Mompox y Loba.* Carlos Valencia Editores. Bogotá. 1979.

FAJARDO, Darío. *El Regímen de la Encomienda en la Provincia de Vélez.* ("The Regime of the *Encomienda* in the Province of Vélez"). Universidad de los Andes. Bogotá. 1969.

FRAGOSO PIRES, Fernando Tasso. *Fazendas. The Great Houses and Plantatios of Brazil.* Ed. Abbeville Press. New York. 1995.

GONZALEZ, Margarita. *El Resguardo en el Nuevo Reino de Granada.* ("The *Resguardo* in the New Kingdom of Granada"). La Carreta Inéditos. Bogotá. 1979.

LEYES DE INDIAS. (LAWS OF THE INDIES). *Transcripción de las Ordenanzas de Descubrimiento, Nueva Población y Pacificación* de las *Indias dadas por Felipe II el 13 Julio de 1573, según el original en el Archivo General de Indias de Sevilla.* ("Transcription of the Ordinances of the Discovery, New Settlement and Pacification of the Indies issued by Philip II on the 13th of July 1573, according to the original in the General Archive of the Indies in Seville"). Ed. Ministerio de la Vivienda. Madrid. 1979.

MARTINEZ, Carlos. *Apuntes Sobre el Urbanismo en el Nuevo Reino de Granada.* ("Notes on Urbanism in the New Kingdom of Granada"). Ed. Banco de la República. Bogotá. 1964.

ORTEGA Y GASSET, José. *Teoría de Andalucía.* ("Theory of Andalusia"). Obras Completas. Vol. VI. Ed. Revista de Occidente. Madrid. 1961.

PARDO UMAÑA, Camilo. *Haciendas de la Sabana de Bogotá.* ("Haciendas of the Savannah of Bogotá"). Villegas Editores. Bogotá. 1988.

RENDON GARCINI, Ricardo. *Las Haciendas de México.* ("The Haciendas of Mexico"). Ed. Fomento Cultural Banamex. Ed. Fomento Cultural Banamex. México. D.F. 1994.

RUEDA VARGAS, Tomás. *Escritos.* ("Writings"). Vol. II. *La Sabana.* Antares. Bogotá. 1963.

SAENZ Quesada, María. *Estancias. The Great Houses and Ranches of Argentina.* Ed. Abbeville Press. New York. USA. 1992.

SALCEDO, Jaime. *Guadalajara de Buga y su Arquitectura.* ("Guadalajara de Buga and its Architecture"). No. 19 Revista del Instituto de Investigaciones Estéticas Carlos Arbeláez Camacho de la Universidad Javeriana. Bogotá. 1982.

SARTOR, Mario. *La vivienda mediterránea y la tipología de la casa colonial americana* in *Estudios sobre arquitectura iberoamericana.* Ed. Consejería sobre Cultura y Medio Ambiente. Junta de Andalucía. Sevilla. 1990.

ST. EXUPERY, Antoine de. *Citadelle.* In *Oeuvre Compléte.* Ed. NRF Collection La Pléiade. Paris. 1947.

TELLEZ, Germán. *Casa Colonial.* ("The Colonial House"). Villegas Editores. Bogotá. 1995. *La Casa de Hacienda* ("The Hacienda House"). Chapter in *Historia del Arte Colombiano.* ("History of Colombian Art"). Ed. Salvat, Barcelona, España. 1975. Republished in *Crítica e Imagen* ("Criticism and Image"). ed. Escala. Bogotá. 1977. *Las Casas de Hacienda Coloniales en Colombia* ("Colonial Hacienda Houses in Colombia"), in *Estudios Sobre Arquitectura Iberoamericana.* Ed. Consejería sobre Cultura y Medio Ambiente, Junta de Andalucía, Sevilla. 1990. *Cultura y Technología en la Arquitectura Colonial de la Nueva Granada* ("Culture and Technology in the Colonial Architecture of Nueva Granada"), lecture in the seminar *Influencias Hispánicas en la Arquitectura Americana,* ("Spanish Influences on American Architecture"), Columbia University, New York, USA, 1985.

TERAN Bonilla, José Antonio. *Arquitectura Rural en México. Las Haciendas de Una Región.* ("Rural Architecture in Mexico. The Haciendas of a Region"). In *Estudios sobre Arquitectura Iberoamericana.* Ed. Consejería para la Cultura y el Medio Ambiente. Junta de Andalucía. Sevilla. 1990.

TOBON, Néstor. *La Arquitectura de la Colonización Antioqueña.* ("The Architecture of the Antioqueña Colonization"). 5 vols. Villegas Editores. Bogotá, 1980-85.

WOBESER, Gisela von. *La Formación de la Hacienda en la Epoca Colonial. El uso de la tierra y del agua.* ("The Formation of the Hacienda in the Colonial Epoch. The use of land and water"). Universidad Autónoma de México. 1983.

Index of photographs

ANTONIO CASTAÑEDA, Jacket, End papers, Pages: 1, 2, 4/5, 6, 8, 16, 17, 38, 39, 40a, b, 41, 44b, 45, 46, 48, 49, 58, 62, 63, 64a, b, 65, 70, 71, 77a, 80/81, 82, 83, 84a, b, 85, 86, 87a, b, 101, 102, 104, 105a, b, 109, 112a, b, 113, 114, 115a, b, 116/117, 118, 119, 120, 121, 122, 123, 126, 127, 128, 129, 152a, b, 153, 154a, b, 155, 158a, b, 159, 160/161, 162, 163a, b, 166, 167, 168a, b, 169, 170, 171, 172, 173b, 174, 175, 176a, b, 177a, b, 178, 179, 180, 181, 182a, b, 183, 184/185, 200, 202/203, 204a, 205, 207, 208a, 210, 211, 212, 213, 214/215, 216/217, 218, 219, 220/221, 222/223, 224a, 225, 226a, 227, 232a, b, 233, 234/235, 236 237a, b, 238, 239, 240a, b, 241, 242, 243, 246, 247, 248a, b, 249, 250, 251, 252, 253, 254, 255a, b, 256a, b, 257, 258, 259a, 260, 261a, b, 262/263, 264, 265, 269, 270a, b, 271, 276a, b, 277, 278a, b, 279, 280, 281, 283, 284a, b, 285, 286, 287, 288a, b, 289, 294, 295, 296, 300.

CLAUDIA URIBE, 18/19, 20a, b, 21a, b, 34, 35a, b, 36, 37, 106, 108, 124, 125a, b, 186b, 196, 208b, 209, 224b, 226b, 230a, b, 231, 259b, 266a, b, 267, 268, 276a, 282, 290a, b, 291, 292, 293a, b.

GERMAN TELLEZ, 10, 14/15, 22, 26, 50, 54, 66a, 72, 73a, b, 74a, b, 75, 76, 77b, 78, 79, 88, 92a, b, 100, 103, 132, 136, 140/141, 142a, b, 143, 144/145, 149, 164, 165, 173a, 192, 204b, 206, 273, 274.

CARLOS HUMBERTO ARANGO, 156a, b, 157, 186a, 187a, b, 190, 191a, b, 244, 245a, b.

JEREMY HORNER, 42a, b, 43, 44a, 66b, 67, 68a, b, 69.

JUAN CARLOS BALCAZAR, 188a, b, 189, 228, 229a, b.

MIGUEL SALAZAR, 47, 130/131, 147, 148, 150/151.

PILAR GOMEZ, 107, 110/111.

JOSE FERNANDO MACHADO, 30, 96.

IVAN ORTIZ, 272, 275.

DIEGO MIGUEL GARCES, 146.

SILVIA PATIÑO, Back Jacket.